Springer Proceedings in Energy

The series Springer Proceedings in Energy covers a broad range of multidisciplinary subjects in those research fields closely related to present and future forms of energy as a resource for human societies. Typically based on material presented at conferences, workshops and similar scientific meetings, volumes published in this series will constitute comprehensive state-of-the-art references on energy-related science and technology studies. The subjects of these conferences will fall typically within these broad categories:

- Energy Efficiency
- Fossil Fuels
- Nuclear Energy
- Policy, Economics, Management & Transport
- Renewable and Green Energy
- Systems, Storage and Harvesting
- Materials for Energy

eBook Volumes in the Springer Proceedings in Energy will be available online in the world's most extensive eBook collection, as part of the Springer Energy eBook Collection. Please send your proposals/inquiry to Dr. Loyola DSilva, Senior Publishing Editor, Springer (loyola.dsilva@springer.com)

More information about this series at http://www.springer.com/series/13370

Iosif Mporas · Pandelis Kourtessis ·
Amin Al-Habaibeh · Abhishek Asthana ·
Vladimir Vukovic · John Senior
Editors

Energy and Sustainable Futures

Proceedings of 2nd ICESF 2020

 Springer

Editors
Iosif Mporas
University of Hertfordshire
Hatfield, UK

Pandelis Kourtessis
University of Hertfordshire
Hatfield, UK

Amin Al-Habaibeh
Nottingham Trent University
Nottingham, UK

Abhishek Asthana
Sheffield Hallam University
Sheffield, UK

Vladimir Vukovic
Teesside University
Middlesbrough, UK

John Senior
University of Hertfordshire
Hatfield, UK

ISSN 2352-2534 ISSN 2352-2542 (electronic)
Springer Proceedings in Energy
ISBN 978-3-030-63918-1 ISBN 978-3-030-63916-7 (eBook)
https://doi.org/10.1007/978-3-030-63916-7

This Springer imprint is published by the registered company Springer Nature Switzerland AG
The registered company address is: Gewerbestrasse 11, 6330 Cham, Switzerland

Preface

Ensuring future security and sustainability of energy supply and the management of energy demand has become a global challenge. Whether the communities of the future are mega-cities, urban, rural or remote, the infrastructure that supports them—including housing, manufacturing, transport, and services—will rely on an energy generation and distribution network that is stable, secure, and resilient to climate change. This network will need to be sustainable in both environmental and financial terms, integrate with more intelligent systems, and be developed alongside more effective policies.

The 2nd International Conference on Energy and Sustainable Futures (ICESF 2020), organized by University of Hertfordshire and Doctoral Training Alliance (DTA) in Energy[1] of University Alliance, was intended for experts in the field from industry and academia as well as research students and early career researchers.

This book contains a collection of submitted papers presented at the ICESF 2020 conference, which were thoroughly reviewed by members of the Program Committee consisting of 22 top specialists in the conference topic areas. A total of 34 papers were selected by the Technical Program Committee for the presentation at the conference and for inclusion in this book. Theoretical and more general contributions were presented in online sessions, due to the COVID-19 pandemic. Topic-specific sessions as well as keynote talks from Dr. John Vardakas/Prof. Christos Verikoukis (Iquadrat snd CTTC/CERCA) and Mr. Gerard Barron (DeepGreen), and panel discussions, brought together specialists in limited problem areas with the aim of exchanging knowledge resulting from research works and previous experience.

ICESF 2020 was organized in six major thematic areas:

- Energy Storage and Sources
- ICT and Control
- Renewables
- Electric Vehicles and Transportation Technology
- Energy Governance, Policy, and Sustainability
- Materials

---

[1] https://unialliance.ac.uk/dta/programmes/dta-energy/.

This publication was supported by the European Commission's Horizon 2020 DTA3/COFUND[2] project under the Marie Skłodowska-Curie grant agreement No. 801604.

Hatfield, UK   Iosif Mporas
Hatfield, UK   Pandelis Kourtessis
Nottingham, UK   Amin Al-Habaibeh
Sheffield, UK   Abhishek Asthana
Middlesbrough, UK   Vladimir Vukovic
Hatfield, UK   John Senior
November 2020

---

[2]https://unialliance.ac.uk/dta/cofund/.

Acknowledgements

We would like to express our gratitude to the authors for providing their papers on time, to the Members of the ICESF 2020 Technical Program Committee for their careful reviews, and to the editors and correctors for their hard work in preparing this book. Special thanks are due to the members of the Local Organizing Committee, Vahid Hosseini, Surya Maruthupandian, and Pascal Schirmer, for their tireless effort and enthusiasm during the conference organization as well as to the Head of the Doctoral Training Programmes of University Alliance, Jennie Eldridge, for her contribution and help in the conference preparation, organization, and publication of this book.

Hatfield, UK	Iosif Mporas
Hatfield, UK	Pandelis Kourtessis
Nottingham, UK	Amin Al-Habaibeh
Sheffield, UK	Abhishek Asthana
Middlesbrough, UK	Vladimir Vukovic
Hatfield, UK	John Senior
November 2020	

Contents

Part I
Energy Storage and Sources

Chapter 1
Solute Driven Transient Convection in Layered Porous Media

Emmanuel E. Luther, Seyed M. Shariatipour, Michael C. Dallaston, and Ran Holtzman

Abstract CO_2 geological sequestration has been proposed as a climate change mitigation strategy that can contribute towards meeting the Paris Agreement. A key process on which successful injection of CO_2 into deep saline aquifer relies on is the dissolution of CO_2 in brine. CO_2 dissolution improves storage security and reduces risk of leakage by (i) removing the CO_2 from a highly mobile fluid phase and (ii) triggering gravity-induced convective instability which accelerates the downward migration of dissolved CO_2. Our understanding of CO_2 density-driven convection in geologic media is limited. Studies on transient convective instability are mostly in homogeneous systems or in systems with heterogeneity in the form of random permeability distribution or dispersed impermeable barriers. However, layering which exist naturally in sedimentary geological formations has not received much research attention on transient convection. Therefore, we investigate the role of layering on the onset time of convective instability and on the flow pattern beyond the onset time during CO_2 storage. We find that while layering has no significant effect on the onset time, it has an impact on the CO_2 flux. Our findings suggest that detailed reservoir characterisation is required to forecast the ability of a formation to sequester CO_2.

Keywords Layered heterogeneity · Convection · Numerical simulations · CO_2 storage · Climate change

1.1 Introduction

The behaviour of CO_2 in a brine-saturated porous medium during CO_2 geological storage involves several processes. According to IPCC [1], a combination of traps safely store injected CO_2 within an underground storage formation. The vertical migration of buoyant CO_2 is restricted to lateral spreading by the formation seal

E. E. Luther (✉) · S. M. Shariatipour · R. Holtzman
Fluid and Complex Systems Research Centre, Coventry University, Coventry, UK
e-mail: luthere@coventry.ac.uk

M. C. Dallaston
School of Mathematical Sciences, Queensland University of Technology, Brisbane, Australia

© The Author(s) 2021
I. Mporas et al. (eds.), *Energy and Sustainable Futures*, Springer Proceedings in Energy,
https://doi.org/10.1007/978-3-030-63916-7_1

3

(structural trap), the imbibing resident fluid can cut-off and immobilize trailing CO_2 plume within formation pores (residual trap), the dissolution of CO_2 in resident fluid removes CO_2 from a highly mobile phase (solubility trap), and the reaction of CO_2 saturated brine with formation rock securely stores CO_2 (mineralization trap). Tracking the buoyant migration, spreading, dissolution, and reaction of CO_2 simultaneously is non-trivial since they occur at different temporal and spatial scales. However, understanding the subsurface behaviour of CO_2 is crucial to controlling the global climate change through safe implementation of CO_2 geological sequestration.

To overcome the modelling challenges, many researchers focused on one or more of the trapping mechanisms at a time [2–5]. Our interest is on solubility trapping which progresses to trigger density driven convection. CO_2 dissolution rate, which measures solubility trapping, has been estimated for Sleipner field in Norway [6] while downwelling convective fingers are visually observed in experimental studies on convection in Hele-Shaw cells [7]. Previous studies of convection in homogeneous systems use the onset time when the instability commences and the critical wavelength of the convective fingers to characterise convection [2, 3]. The flow pattern after the onset time can be described in several regimes: diffusive, flux growth when the convective instabilities grow, constant flux when the fingers progress towards the bottom of the domain, and final period (when flux gradually declines) [4]. In heterogeneous systems, many studies on transient convection are in systems with random permeability field or impermeable dispersed barriers [5]. Layering which exist naturally in sedimentary rocks has not received worthy attention, but it can affect the dynamics of fluid flow due to the change in permeability. Therefore, we use numerical simulations with perturbations from numerical artefacts to show that layered systems can alter CO_2 flux beyond the onset time.

1.2   Governing Equations and Model

$$\frac{\partial u}{\partial x} + \frac{\partial v}{\partial z} = 0 \tag{1.1}$$

$$u = -\frac{k}{\mu}\frac{\partial p}{\partial x} \tag{1.2}$$

This problem of density driven convection in a brine saturated porous medium is modelled in single phase comprising the flow of brine and the transport of CO_2 dissolved in brine. Each model aquifer is 2-dimensional in the cartesian coordinate system (x-z plane). The aquifer has a thickness H, impermeable $(u = v = 0)$ and constant CO_2 concentration $(c = C_s)$ at the top boundary $(z = 0)$, and impermeable and zero CO_2 concentration flux $\left(\frac{dc}{dz} = 0\right)$ at the bottom boundary $(z = H)$. Initially $(t = 0)$, the domain is assumed to contain no CO_2 $(c = 0)$. This initial and boundary conditions are common in the subject of convection in the context of the geological sequestration of CO_2 [2–5]. The system of equations for this problem is

$$v = -\frac{k}{up\mu}\left(\frac{\partial p}{\partial z} - (1 + \beta c)\rho_0 g\right) \tag{1.3}$$

$$\emptyset\frac{\partial c}{\partial t} + u\frac{\partial c}{\partial x} + v\frac{\partial c}{\partial z} = \emptyset D\left(\frac{\partial^2 c}{\partial x^2} + \frac{\partial^2 c}{\partial z^2}\right) \tag{1.4}$$

$$v = -\frac{k}{\mu}\left(\frac{\partial p}{\partial z} - (1 + \beta c)\rho_0 g\right) \tag{1.5}$$

where c, k, D, $\emptyset$, g, μ and p denote CO_2 dissolved concentration, permeability, diffusivity of CO_2 in brine saturated porous medium, porosity, gravity, viscosity and pressure respectively. The Darcy velocity components are u in the horizontal and v in the vertical direction. Boussinesq approximation is made so that the dependence of density (ρ) on the concentration (c) which is assumed to be linear $\rho = \rho_0(1 + \beta c)$ is applied only in the buoyancy term (ρg) in (3), where ρ_0 is the density of pure brine and β is the expansion coefficient. The length, time, and concentration scale are H, the diffusive timescale $\frac{H^2}{D}$, and C_s respectively. The velocity scale is $\frac{\emptyset D}{H}$, and the dimensionless pressure is expressed as $p' = \frac{(p - \rho_0 g z)}{\emptyset\mu D}$. The main dimensionless number that describes the relative importance of diffusion and convection is the Rayleigh number, $Ra = \frac{k\rho_0\beta C_s g H}{\emptyset\mu D}$. The system of Eqs. (1.1–1.4) is made dimensionless, linearized, and then analysed using linear stability analysis (LSA) based on the quasi-steady state approximation (QSSA) needed due to the transient base profile. The procedure and validity of QSSA have been discussed previously [3].

1.3   Numerical Simulations

The governing equations and boundary conditions are numerically solved using Eclipse 300 which is a multi-phase, compositional reservoir simulator. Each model is 2 m × 1.01 m in the horizontal (x) and vertical (z) directions respectively discretized into 200 grids in x ($i = 1, 2 \ldots 200$) and 101 grids in z ($j = 1, 2 \ldots 101$) directions, where i and j are the spatial discretizations. The initial and boundary conditions are as described in Sect. 1.2. We rely on the numerical artefacts to trigger the instability.

1.4   Results and Discussion

The relationship $t'_0 = a Ra^{-2}$ between the onset time (t'_o) and the Rayleigh number in the homogeneous systems is obtained both in the LSA and in the numerical simulations (Fig. 1.1a). This relationship and the constant $a \approx 56$ from the LSA in this work are similar to that obtained in the previous study [3]. However, in comparison to the theoretical result, the numerical result for the constant $a \approx 807$ is large due to weak perturbations from numerical artefacts which can delay the onset time. This

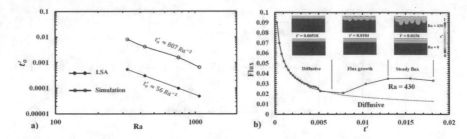

Fig. 1.1 a LSA and numerical simulation results for the relationship between the onset time and Rayleigh number in homogeneous systems for Ra = 322.97, 430.44, 751.87, 1070.63. **b** Average flux vs time plot for Ra = 0 (diffusive system) and Ra = 430.44 also showing the profile for CO_2 liquid mole fraction (c') at the top. The variation in Ra is solely due to permeability difference. Both plots are in dimensionless unit. The onset time in the LSA is the time when the perturbations growth rate become positive while the onset time in the simulation is when the standard deviation of the horizontal concentration profile across the domain increases

can also be due to the difference in the definition of the onset time in the LSA and simulation (Fig. 1.1). Beyond the onset time, LSA breaks down, and we depend on the numerical simulations to observe the flow regimes. We estimate the average flux (*F*) using the rate of change of the total dissolved concentration (CO_2 liquid mole fraction) in the domain given by $F = \frac{1}{2}\frac{d}{dt}\left(\int_0^2\int_0^1 cdxdz\right)$ which is mathematically equivalent to the average flux from the top boundary $\left(-\frac{1}{2}\int_0^2\frac{dc}{dz}dx\right)$. The diffusive, flux growth and steady flux regimes that we find (Fig. 1.1b) have been previously observed [4]. The scaling agreement between our numerical simulation and LSA results for the onset time, and the similarity of our simulation results with previous numerical studies for the flow regimes after the onset time validates our numerical models. Hence, we extend this analysis to include heterogeneous cases using our simulation models.

The heterogeneous models each has a depth, *H*, and comprises two homogeneous media with one overlaying the other. The thickness of each of the homogeneous media is half of the depth of the full system $\left(\frac{1}{2}H\right)$. We investigate the effect of the layered heterogeneity on the onset time of convective instability and on the subsequent flow pattern. The nomenclature, TDiff - B430, interpreted as a heterogeneous system in which the top strata is a diffusive system (Ra = 0) and the bottom layer has Ra = 430.44 is adopted.

The onset of convective instability is not affected by the layering (Fig. 1.2). The top layer determines the onset time, and the bottom layer has no role in this setting. Similar to the onset time, the time of the flux growth depends largely on the properties of the top layer (Fig. 1.3). The time of flux growth in the heterogeneous systems with Ra = 430.44 at the top are similar and the amplitude of the growth can be approximated by that in the homogeneous system with Ra = 430.44. However, after

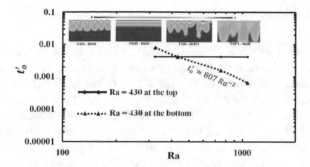

Fig. 1.2 Numerical simulation results for the relationship between the onset time and Rayleigh number in heterogeneous systems for Ra = 322.97, 430.44, 751.87, 1070.63 when Ra = 430.44 is at the top and at the bottom. At the top of this figure, the profile of CO_2 mol fraction in liquid phase in T430-BDiff, TDiff-B430, T430-B1071 and T1071-B430 at $t' = 0.018$ are presented

Fig. 1.3 Average flux vs time plot for the homogeneous case Ra = 430.44 and the presented heterogeneous cases

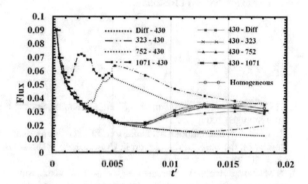

the flux growth, the flux profile depends on the Ra of the bottom layer when dissolved CO_2 reaches this layer. The system with the purely diffusive bottom layer has a sharp flux decline (Fig. 1.3) due to the restriction of flow from the permeable top layer by the diffusive bottom layer (see concentration profile in Fig. 1.2). When a high Ra layer lies above a low Ra layer (T1071-B430 and T751- B430), the width of the convective fingers reduces when the fingers reaches the bottom layer reducing the flux (Fig. 1.2). The flux growth is controlled by the permeability in the top region, and during late times, the flux can be approximated by the flux in the homogeneous system Ra = 430.44. These results suggest that we can fairly estimate the flux in some layered heterogeneous systems with the flux in homogeneous media at certain times.

1.5   Conclusions

The role of layered permeability heterogeneity on CO_2 solute convection in a brine saturated geological porous medium has previously not received much attention. We developed a numerical model on Eclipse 300 and validated our numerical results against theoretical work for the onset of convective instability and against previous studies for the flow regime beyond the onset time. Our results indicate that the bottom layer has no significant effect on the onset of convective instability while dissolved CO_2 remains in the top layer. Beyond the onset time, the presence of the bottom layer affects the flow regimes. This result implies detailed knowledge of potential storage formations is required to successfully implement CO_2 storage.

Acknowledgements This work is produced through the sponsorship of Petroleum Technology Development Fund (PTDF) in Nigeria and is supported by the Centre for Fluid and Complex Systems, Coventry University, UK. We also acknowledge Schlumberger for the use of Eclipse 300 and Amarile for the use of Re-Studio.

References

1. IPCC (Intergovernmental Panel on Climate Change), Special Report on Carbon Dioxide Capture and Storage (2005)
2. E. Luther, S. Shariatipour, M. Dallaston, Effect of numerical error due to heterogeneous permeability on the onset of convection in CO_2 storage. 81st EAGE Conference and Exhibition (1), 1–5 (2019)
3. S.M. Jafari Raad, H. Hassanzadeh, 'Onset of dissolution-driven instabilities in fluids with nonmonotonic density profile. Phys. Rev. E, Stat. Nonlin. Soft Matter Phys. **92**(5), 053023 (2015)
4. K. Pruess, K. Zhang, 'Numerical modeling studies of the dissolution-diffusion-convection process during CO_2 storage in saline aquifers. Lawrence Berkeley National Lab. (LBNL), Berkeley, CA (United States) (2008)
5. C.P. Green, J. Ennis-King, Steady dissolution rate due to convective mixing in anisotropic porous media. Adv. Water Resour. **73**, 65–73 (2014)
6. A. Furre, O. Eiken, H. Alnes, N.J. Vevatne, F.A. Kiaer, 20 years of monitoring CO_2—injection at Sleipner. Energy Procedia **114**, 3916–3926 (2017)
7. J.T. Kneafsey, K. Pruess, Laboratory Flow Experiments for Visualizing Carbon Dioxide-Induced, Density-Density Brine Convection. Transp. Porous Med. **82**, 123–139 (2010)

Chapter 2
Towards an Optimal Deep Neural Network for SOC Estimation of Electric-Vehicle Lithium-Ion Battery Cells

Muhammad Anjum, Moizzah Asif, and Jonathan Williams

Abstract This paper has identified a minimal configuration of a DNN architecture and hyperparameter settings to effectively estimate SOC of EV battery cells. The results from the experimental work has shown that a minimal configuration of hidden layers and neurons can reduce the computational cost and resources required without compromising the performance. This is further supported by the number of epochs taken to train the best DNN SOC estimations model. Hence, demonstrating that, the risk of overfitting estimation models to training datasets, can also be subsided. This is further supported by the generalisation capability of the best model demonstrated through the decrease in error metrics values from test phase to those in validation phase.

Keywords Deep neural network (DNN) · State of charge (SOC) · Lithium-ion · EV · Energy storage

2.1 Introduction

Electric Vehicles (EV's) and Hybrid Electric Vehicle (HEV's) are rapidly becoming an essential mode of modern transportation. The research on green energy and transport systems will lead to the adoption of EVs widely across the globe. Driving a battery powered vehicle does not generate harmful exhaust fumes as opposed to vehicles which run on gasoline. Hence, studies on developing effective capacity estimation methods for battery systems of EVs play a critical role in the improvement of EV's and HEV's battery life and range prediction research [1, 2]. Li-ion batteries are preferred over other battery chemistries, due to certain advantages such as high energy density [3, 4]. However, like any other cell chemistry, Li-ion also comes with certain common limitations, such as specific operational temperature and voltage

M. Anjum (✉) · J. Williams
Centre for Automotive & Power Systems Engineering, University of South Wales, Pontypridd, UK
e-mail: muhammad.anjum@southwales.ac.uk

M. Asif
School of Computing and Mathematics, University of South Wales, Pontypridd, UK

© The Author(s) 2021
I. Mporas et al. (eds.), *Energy and Sustainable Futures*, Springer Proceedings in Energy,
https://doi.org/10.1007/978-3-030-63916-7_2

11

range, and a need to improve the accuracy of capacity estimation. Likewise, a range of external and internal factors, also influences Li-ion battery's performance. These may include temperature, voltage, charge/discharge cycles, ageing process, depth of discharge (DOD), and battery's internal electro-chemical processes. Consequently, all the above stated factors affect the stabilisation of an EV's battery pack and estimation of its driving range. Hence, an optimal framework for an EV battery capacity estimation is crucial to ensure reliable operation of battery management [2, 5, 6].

2.2 Background and Motivation

Estimating a battery's capacity state also provides measures to assess its performance, health, and monitor the ageing process. A battery is easily influenced by the variation in external factors values due to vehicle load and driving habits. These driving habits may include, but are not limited to variation in the speed and braking. Such driving behaviours render capacity estimation of EV battery more challenging. Furthermore, the capacity estimation of an EV has to account for the factors mentioned in Sect. 2.1, as they affect battery performance [5, 6]. One of the measures to assess the battery capacity is the estimation of its state of charge (SOC), which is the ratio between the outstanding capacity of a battery to its available capacity [7]. A mathematical representation of SOC is shown below in Eq. (2.1) where Q_n is the available capacity of the battery and Q_m is the outstanding capacity of the battery.

$$SOC\% = \frac{Q_n}{Q_m} \times 100\% \tag{2.1}$$

Several methods have been used by researchers to estimate SOC, such as conventional methods, adaptive filter algorithms and hybrid methods as categorised in our previous work [8]. All these methods come with some drawbacks for estimating EV battery's SOC. These drawbacks include ruling out the effects of variation in external factors such as current, voltage and temperature on battery behaviour and life.

2.2.1 Machine Learning Methods for SOC Estimation of EV Batteries

Besides the aforementioned methods, machine learning algorithms, especially deep neural networks (DNN), have been widely used and proposed to estimate SOC of EV batteries. Some researchers have integrated other methods such as Kalman filters with equivalent circuit battery models to extract battery parameters in conjunction with DNNs to estimate SOC [9]. Whereas, others have extracted battery parameters by using battery testers and estimated SOC using DNNs [10, 11]. Works, such as Du.

et al. have used constant temperature and steady discharge pulses for battery parameter extraction [12]. The extraction of steady state condition input parameters for
learning algorithms, renders the learned model untrained for transient load demand
and real-world situations.

On the other hand, works including and similar to Chemali et al.'s, have extracted
data from non-steady state conditions by simulating different custom and standard-
ised drive cycles, such as UDDS at multiple temperatures and C-rate [11]. However,
to the best of authors' knowledge, such works run the risk of overfitting the proposed
models to training data by running tens of thousands of learning epochs and iterations
to minimise the error metrics. Hence, possibly resulting in significant computational
time and memory resources.

Following the discussion in this and the previous section, two research questions
have been formulated:

*RQ1—can the SOC of Li-ion EV cells be effectively estimated by identifying a
minimal DNN architecture (number of hidden layers and neurons)?*
*RQ2—can the number of learning epochs of the DNN architecture proposed in
RQ1 be optimised by searching the DNN hyperparameters space to estimate the
SOC of Li-ion EV cells?*

2.3   Experimental Setup and Design

The experiment was designed into four main phases, as shown in Fig. 2.1. The first
phase is to facilitate model parameter extraction, which would be used as inputs
to the DNNs. The second phase is to train a number of DNNs over a range of
hyperparameter settings. Each DNNs would be validated to find the optimal model
for SOC estimation with the given set of training input. The last two phases deal with
testing and evaluating the best model determined from validation in the previous
phase against test data. The performance is measured using multiple error metrics,
followed by estimated result's analysis to draw conclusion and recommendations for
future work.

2.3.1   Data Extraction

Capacity (CAP) tests via a battery cell tester (Bitrode) were conducted on eight 20
Ah LiFePO$_4$ cells at a range of temperature variations $T/°C = [0, 10, 25, 35, 45]$

Fig. 2.1 Experimental phases

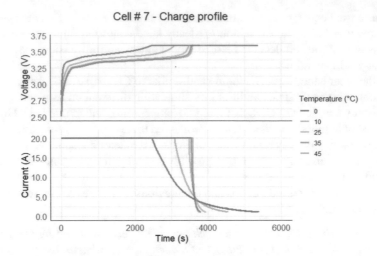

Fig. 2.2 CAP test charge profile of a LiFePO$_4$ cell on various temperatures

per cell, at the CAPSE Laboratory, University of South Wales. A constant current of 1 C-rate was applied to charge and discharge cells in the battery tester; along with an environmental chamber to maintain the required temperatures, respectively. The sampling rate was set to be 0.1 Hz and the data extracted from all of the 40 CAP tests was individually stored on a host computer. The parameters extracted for DNN input included voltage: V/V (voltage), current: C/A, both battery and external temperature via environmental chamber's temperatures in degree Celsius: T_b and T_e, charged capacity of the cell in ampere hour: Q_n/Ah, and battery state: S (rest, charge, and discharge). A typical charging profile based on the model parameters extracted from one of the cells is shown in Fig. 2.2. The effect of various temperatures on a LiFePO$_4$ cell's behaviour during the CAP test around voltage and current can be clearly seen from the plots and reinforces the need to train estimation models on a range of temperatures.

2.3.2 Deep Learning for SOC Estimation

The learning phase for SOC estimation was carefully designed to find answers to the research questions posed in Sect. 2.2. The aim was to find an optimal trend of hyperparameter values for a minimally configured DNN. Thus, achieving an effective SOC estimation model with minimal values of error metrics. Consequently, the study also focused on alleviating the need for using high specification hardware, such as GPUs for the learning phase by finding a trend of minimal number of neurons as well as hidden layer configuration for DNN architecture. A greedy search approach was deployed to find the optimal number of hidden layers and neurons along with other

Table 2.1 DNN hyperparameters list for greedy search

Hyperparameter	Values
Hidden layers and neurons	[5, 7, 10], [15, 20][a]
Learning rate	0.5, 0.8
Activation method	Hyperbolic, rectifier, and maxout
Epochs	10, 25, 50, 75, 100

[a]The number of elements in a set represent the number of hidden layers, whereas the value of each element represents the number of neurons in that hidden layer. Each configuration is separated by a comma and enclosed in set notation

DNN hyperparameters values. Some of the hyperparameter and DNN configurations searched are listed in Table 2.1. All possible DNN architectures and hyperparameter settings resulted in 360 DNN models. Each model was trained and cross validated with 5 folds, using the entire data extracted from 6 out of 8 cells. Whereas 2 cells' data was used to test the models, which approximately accounts for 25% of the extracted data.

2.4 Results and Discussion

To evaluate the DNNs performance, using the error metrics value between actual SOC and estimated SOC; actual SOC was calculated using the formula provided in (1). The top 20 models, determined from the validation phase, shared common DNN architecture: [10,15,20] and hyperparameter settings except for the number of epochs. The learning rate for the top 10 models was 0.8, which signifies the generalisation capability of the architecture, as compared to the other smaller values provided to greedy search. The linear and locally weighted regression smoothing curves in the epochs vs residual error plot in Fig. 2.3: *Actual and Predicted SOC from best estimation models at a range of temperatures [0,10,25,35,45] °C*

Fig. 2.3 Actual and Predicted SOC from best estimation models at a range of temperatures [0,10,25,35,45] °C

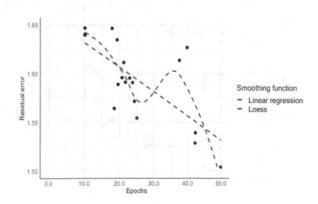

Figure 2.4 verify the decreasing trend in error rate as the DNN SOC estimation models' learning epoch increased. Another interesting observation can be made from this plot, where each point in the plot shows the residual error at its final epoch. A 40% decrease ~(50–10) in number of epochs has cost a less than ~0.15 increase in error metrics values. If the trend in the cost vs number of epochs remains similar with larger and richer training datasets, a slight trade-off on accuracy can bring significant reduction in temporal costs and computing resources in the learning phase of SOC estimation. A number of error metrics were used to evaluate the performance of the best model other than residual error as shown in Table 2.2. The decrease in each error metrics' value from validation to test phase provides evidence for the model's generalisation capability as it has performed better on unseen test data. The actual and estimated SOC plots at charging from the best model in in Fig. 2.4 also depict that the model has generalised well on unseen data and did not overfit on training data. This is further supported by the low MAE reported in Table 2.2. as well as its comparison with Chemali et al.'s MAE values on three different test data sets: 0.39, 1.85 and 1.35 [12]. It is worth mentioning that while the lowest MAE is lower and highest MAE is higher than the reported MAE in this work, the number of learning epochs for their work (85000) is significantly higher than the best model's learning epochs (~50).

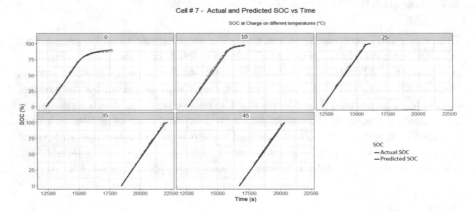

Fig. 2.4 Epoch vs error along with smoothed trends curves

Table 2.2 Error metrics of the best SOC estimation model from validation and test phase

Error metric	Validation	Test
Mean squared error (MSE)	26.41	24.05
Root mean squared error (RMSE)	5.14	4.90
Mean absolute error (MAE)	1.68	1.39

2.5   Conclusion

Based on the analysis driven from results and the discussion in the previous section we can answer the research questions posed in Sect. 2.2. A minimal DNN architecture has been identified from 360 DNN models with different configurations of minimal architectures to effectively estimate EV battery cells SOC with a performance metrics value on par with state of the art research work. Furthermore, the number of learning epochs to reduce computation resources and temporal costs, as well as to avoid risk of overfitting, have shown to be significantly less in comparison with existing research works to the best of the authors' knowledge.

The DNN estimation models can be further improved by training and testing on data extracted from more battery tests, such as HPPC tests on a range of temperatures and C-rate. Similarly, adding custom and standard drive cycles data extracted from the battery cells to both the training and testing sets can further inform the estimation model on impacts of external factors affecting EV battery's behaviour.

References

1. D. Linden, Handbook of batteries and fuel cells. *New York, McGraw-Hill Book Co., 1984, 1075 p. No individual items are abstracted in this volume* (1984)
2. F. Herrmann, F. Rothfuss, Introduction to hybrid electric vehicles, battery electric vehicles, and off-road electric vehicles, in *Advances in Battery Technologies for Electric Vehicles* (pp. 3–16). Woodhead Publishing (2015)
3. E. Chemali, M. Preindl, P. Malysz, A. Emadi, Electrochemical and electrostatic energy storage and management systems for electric drive vehicles: state-of-the-art review and future trends. IEEE J. Emerg. Sel. Topics Power Electron. **4**(3), 1117–1134 (2016)
4. V. Pop, H.J. Bergveld, D. Danilov, P.P. Regtien, P.H. Notten, *State-of-the-art of battery state-of-charge determination* (Accurate state-of-charge indication for battery-powered applications, Battery Manag. Syst., 2008), pp. 11–45
5. A. Emadi, *Advanced Electric Drive Vehicles*. CRC Press (2014)
6. M. Li, Li-ion dynamics and state of charge estimation. Renew. Energy **100**, 44–52 (2017)
7. D.Z. Li, W. Wang, F. Ismail, A mutated particle filter technique for system state estimation and battery life prediction. IEEE Trans. Instrum. Meas. **63**(8), 2034–2043 (2014)
8. M. Anjum, K. Thanapalan, J. Williams, Neural network based models for online SOC estimation of LiFePO$_4$ batteries used in electric vehicles, in 2019 ICESF. ICESF (2019)
9. W. Wang, D. Wang, X. Wang, T. Li, R. Ahmed, S. Habibi, A. Emadi, Comparison of Kalman Filter-based state of charge estimation strategies for Li-Ion batteries, in *2016 IEEE Transportation Electrification Conference and Expo (ITEC)* (pp. 1–6). IEEE (2016)
10. C. Bo, B. Zhifeng, C. Binggang, State of charge estimation based on evolutionary neural network. Energy Convers. Manag. **49**(10), 2788–2794 (2008)
11. E. Chemali, P.J. Kollmeyer, M. Preindl, A. Emadi, State-of-charge estimation of Li-ion batteries using deep neural networks: a machine learning approach. J. Power Sources **400**, 242–255 (2018)
12. J. Du, Z. Liu, Y. Wang, State of charge estimation for Li-ion battery based on model from extreme learning machine. Control Eng. Pract. **26**, 11–19 (2014)

Chapter 3
A Brief Review on Nano Phase Change Material-Based Polymer Encapsulation for Thermal Energy Storage Systems

Muhammad Aamer Hayat and Yong Chen

Abstract In recent years, considerable attention has been given to phase change materials (PCMs) that is suggested as a possible medium for thermal energy storage. PCM encapsulation technology is an efficient method of enhancing thermal conductivity and solving problems of corrosion and leakage during a charging process. Moreover, nanoencapsulation of phase change materials with polymer has several benefits as a thermal energy storage media, such as small-scale, high heat transfer efficiency and large specific surface area. However, the lower thermal conductivity (TC) of PCMs hinders the thermal efficiency of the polymer based nano-capsules. This review covers the effect of polymer encapsulation on PCMs while concentrating on providing solutions related to improving the thermal efficiency of system.

Keywords Nano-phase change materials · Polymer encapsulation · Thermal energy storage · Nanotechnology · Heat transfer enhancement

3.1 Introduction

The main factors pushing the world towards the use of renewable energy sources are the continuous increase in carbon emissions and the increase in fuel costs. Direct solar radiations are considered among the most potential source of energy in many parts of the globe. The researcher's community around the globe is looking for renewable and novel energy sources. The storage of energy in suitable forms, which can be converted traditionally into the required form, is a challenge to the technologists of today. Energy storage not only eliminates the difference between demand and supply, but also increases system efficiency and reliability and performs a significant role in energy conservation [1]. The different energy storage techniques are given in Fig. 3.1.

M. A. Hayat · Y. Chen (✉)
Department of Engineering, University of Hertfordshire, Hatfield, Herts AL10 9AB, UK
e-mail: y.k.chen@herts.ac.uk

M. A. Hayat
e-mail: m.hayat2@herts.ac.uk

© The Author(s) 2021   19
I. Mporas et al. (eds.), *Energy and Sustainable Futures*, Springer Proceedings in Energy,
https://doi.org/10.1007/978-3-030-63916-7_3

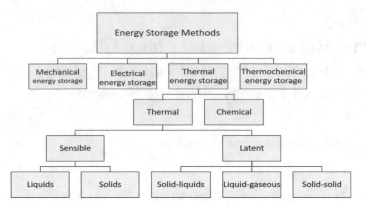

Fig. 3.1 Energy storing techniques [2]

Thermal energy storage (TES) contributes to a significant part in the efficient usage of thermal energy and has utilization in various fields, for instance, in buildings cooling/heating systems, solar collectors, electricity and industrial thermal energy storage [3]. Amongst many thermal energy storage methods, latent heat thermal energy storage is a highly desirable method and has the property of storing heat at a uniform temperature which is the phase change temperature.

Latent heat thermal energy storage (LHTES) which are phase change materials can be classified mainly into two categories i.e. organic and inorganic PCMs. The organic PCMs have higher stability, high energy storage capacity, no segregation, un-toxic, un-corrosive, and un-reactive [4]. Contrary to this, Inorganic PCMs have comparatively higher thermal conductivity, higher density volumetric energy storage, and flame retardance [5]. The organic PCMs have potential advantage as thermal energy storage materials in many applications, such as desalination [6], thermal management of electronic devices [7] passive heating of buildings [8, 9] and other thermal integrated systems. However, PCMs experience less thermal conductivity which is usually (0.2 W m^{-1}K^{-1}) and leakage during the phase transition [10]. The enhancement of thermal conductivity not only increase heat storage and release capacity, but it also improves the performance of the system. There are several methods for the improvement of PCM thermal conductivity, such as by utilizing nanoparticles, encapsulation of PCMs, expanded graphite, fins, heat pipe, and by metallic foams [11].

At present, polymer-based encapsulation of the PCMs attracted the researchers because polymers are flexible which allows the expanded PCM volume during the phase change results in ease of melting while maintaining the stability and shape of the prepared nano-capsules. In addition, encapsulation provides large surface area, high heat transfer rate, prevents leakage and encapsulation also reduces the reactivity of PCMs with external environment. In this study, we will discuss the latest studies on encapsulation of PCMs and its future aspects.

3.2 Polymer Encapsulation-Based Phase Change Materials

Encapsulation is the procedure of enclosing PCMs within coating materials to develop a type of composite PCMs [12]. The major reason to use polymers as core material is that they are mechanically stable, lightweight, inexpensive, and compatible with PCMs [13]. Moreover, the encapsulation technology can be separated into macro encapsulation, micro-encapsulation, and nano-encapsulation, depending on the size. Different forms of physical properties, such as, capillary behaviour, adhesion forces, and surface chemistry, are more efficient at the nanoscale encapsulation. Nano-encapsulation technique has proved to be extra useful than micro and macro-encapsulation for that purpose [14, 15].

The important parameters to evaluate the thermal performance of encapsulated PCMs are core and shell materials, latent heat, melting temperature of PCMs (T_m), encapsulation method and encapsulation efficiency (EE), as listed in Table 3.1.

The data given in Table 3.1 showed that in situ polymerisation techniques exhibit better thermal performance by providing more encapsulation efficiency and thermophysical stability compared to the other encapsulation methods.

Shi et al. [21] examined an interfacial polymerization technique for the development of paraffin-polymethyl methacrylate (PCM-PMMA) nano-capsules. At melting and solidification enthalpy of 64.93 J/g and 66.45 J/g respectively (PCM-PMMA) nano-capsules found stable and reliable. Furthermore, thermal gravimetric analysis (TGA) results showed the decent thermal stability with PCM content of 52.95%. Tumirah et al. [18] experimentally investigated the physical, thermal and chemical properties of the St (styrene)-MMA (methyl methacrylate) copolymer shell with n-octadecane as a core using miniemulsion in situ polymerization. After 360 cycles of heating/cooling, the nano-capsules had reasonable thermal efficiency in terms of chemical stability and thermal properties. The DSC results showed the solidification and melting temperatures of PCMs inside the nano-capsules were 24.6°C and 29.5°C

Table 3.1 Summary of nano-PCMs prepared utilizing various methods

References	Core/PCM	Shell	Latent heat (J/g)	T_m (°C)	EE%	Encapsulation method
[16]	n-octadecane	PMMA/SiO$_2$	178.9	–	10	Sol-gel method
[17]	n-octadecane	PBMA, PBA	96–112	29.1–31.6	47.7–55.6	Suspension-like polymerisation
[18]	n-octadecane	PS-PMMA	107.9	29.5	–	Miniemulsion in situ polymerisation
[19]	n-Dodecanol	Melamine formaldehyde	187.5	21.5	93.1	In situ polymerisation
[20]	n-Nonadecane	PMMA	139.20	31.23	60.3	Emulsion polymerisation

Fig. 3.2 Synthetic
explanation of encapsulated
PCM [24]-reproduced by
permission of The Royal
Society of Chemistry

respectively, which indicates they have a high ability to be utilized for the purpose
of thermal energy storage. Fuensanta et al. [22] studied miniemulsion polymerisation (chemical method) in which (RT-80) PCM with melting temperature 80°C was
utilized as core material and styrene-butyl acrylate copolymer as shell material. The
nano-capsules showed thermal stability even after 200 heating/cooling cycles. In
addition, Differential Scanning Calorimetry (DSC) analysis confirmed the thermal
energy storage capacity of RT80/styrene-butyl acrylate nano-capsules by obtaining
the melting and solidification enthalpies in the range of 10 to 20 J/g. Chen et al. [23]
utilized miniemulsion polymerization method to synthesized styrene-butyl acrylate
(SBA) copolymer as shell and n-dodecanol as core. The thermal performance, particle
size and morphology were measured by DSC, particle size distribution (PSD) and
transmission electron microscope (TEM) respectively. The results revealed that the
encapsulation efficiency (EE) had touched 98.4% and phase transition enthalpy and
phase transition temperature were 10932 J/g and 18.4°C, respectively. Sari et al. [20]
prepared micro/nano capsules by emulsion polymerization method using paraffin
eutectic mixtures (PEMs) as core material and PMMA as shell materials. The TGA
results indicated that the encapsulated PEMs remained durable until 160°C. In addition, after exposure to 5000 thermal cycles, they had good chemical and thermal
stability. A synthetic explanation of encapsulated PCM is shown in Fig. 3.2.

The encapsulation performance is still relatively low and faces the lack of industrial application requirements. What is more, the one reason for its low encapsulation
performance is the very low thermal conductivity of PCM which hinders the heat
transfer rate. Many studies have been investigated in which only PCM is used as the
core material, but rare work is done on improving the thermal conductivity of core
material. The addition of nanoparticles in PCMs increases the TC of PCMs because
they possess high TC materials. We discussed the effects of nanoparticles on the
PCMs in the next section.

3.3 Nanoparticles Based Phase Change Materials (Nano-PCMs)

By increasing the thermal conductivity of PCMs, heat storing and release capacity
surges, which results in the improvement of thermal performance of the system.

In addition, thermal conductivity of the PCMs can be improved by the usage of nanoparticles possessing high thermal conductivity.

Qu et al. [25] studied the impact of two distinct nanoparticles (i.e. Expanded Graphite-Multi-walled Carbon Nano-tube (EG-MWCNT) and Expanded Graphite-Carbon Nano-fiber (EG-CNF)) on the phase change material (Paraffin) at five different mass ratios, and it was found that maximum thermal conductivity increased by the incorporation of EG-MWCNT and EG-CNF was 60% and 21.5% respectively. Rufuss et al. [26] investigated three different nanoparticles (copper oxide (CuO), titanium dioxide (TiO2) and graphene oxide (GO)) with paraffin. The results exhibited that the TC of paraffin was enhanced by 101.2%, 28.8% and 25% by the adding 0.3 wt% of graphene oxide, copper oxide and titanium dioxide nanoparticles, respectively. Sharma et al. [27] experimentally inspected the performance of PCMs and nano-PCMs integrated micro-fins for the Building-Integrated Concentrated Photovoltaics technology. Paraffin wax was used as PCM and Cupric oxide (CuO) as nanoparticles with 0.5% by mass. Results exhibited that the average temperature was decreased by 12.5 °C using micro-fins with nano-PCMs and 10.7 °C using micro-fins with PCMs as comparison to utilizing micro-fins only. Nourani et al. [28] experimentally inspected the effect of Aluminium oxide nanoparticles (Al_2O_3) on paraffin using different concentrations of (Al_2O_3) nanoparticles. The results revealed that the thermal conductivity improvement ratios for liquid and solid states were 13% and 31% respectively for a sample containing 10 wt% of Al_2O_3. Li [29] prepared nano-graphite (NG) and paraffin based composite PCMs. The thermal effects of nano-PCMs were examined using SEM and DSC. The results depict that the thermal conductivity of PCMs increases with the increase in the percentage of nanoparticles. Moreover, addition of 10% of (NG) nanoparticles raised the thermal conductivity to 0.9362 W/m K.

From the literature stated above it is clear that addition of nanoparticles to PCMs improves the thermal conductivity of the PCMs because both metallic and carbon-based nanoparticles have high TC. Moreover, carbon-based nanoparticles, such as carbon nanotubes, carbon fiber and graphene possess better stability, low density, and good dispersion in phase change materials compared to metallic nanoparticles.

3.4   Discussion and Future Work

Polymer-based encapsulated PCMs are widely used in many industrial applications, such as in thermal management, buildings, and medical industry because they have potential to store thermal energy with higher efficiency than other energy storage methods. But still more attention is needed for the further development of the thermal performance of encapsulated PCMs, as suggested below.

- Until now, work was focused on simple PCMs based polymer encapsulation, so future studies need to be conducted on nano-PCMs based polymer encapsulation for the enhancement in the thermal performance of polymer-based nano-capsules.

- In addition, the stability of nano-capsules can be improved by using nano-PCMs as core materials which help in a reduction of encapsulation cost.
- Previously, usually organic PCMs were used as core materials for the micro/nano encapsulation. Hence, there is the need to investigate inorganic PCMs as core materials because they have high latent heat of fusion during phase transition.
- Further studies on improvement of encapsulation efficiency, better thermal performance and better stability need to be conducted.
- Hybrid nanoparticles-based polymer nanocomposite materials also need attention for the development of potential energy storage materials.
- It has been stated that the encapsulation of PCMs results in the reduction of melting temperature latent heat compared to pure PCMs. PCMs aim to use in TES systems as energy storage materials without loss of heat transfer and fluid flow efficiency. This is therefore a major challenge for encapsulated PCMs to raise or sustain the latent fusion heat with different melting and solidification temperatures. Future studies are therefore required to concentrate on encapsulation of PCMs in this direction.

3.5 Conclusion

This paper mainly focused on encapsulation of PCM work success over recent years. Further, addition of nanoparticles in PCMs for the enhancement in the thermal efficiency of polymer-based nano-capsules are also studied. From this study the following findings are summarised.

- PCM encapsulation with a polymer as shell material is easy and does not require any complication, and the introduction of simple polymerisation techniques it can be achieved.
- The problems of leakage, subcooling, and segregation had been somewhat solved after encapsulation of PCMs.
- Addition of high thermal conductive nanoparticles in-to PCMs the thermal performance of encapsulation can be improved.
- In combination with various subsystems such as heat sinks, heat pipes, micro-minichannels, heat exchangers, panels, wallboards, and slabs, encapsulation of PCMs is the most suitable for thermal management and TES applications.

Acknowledgements The authors would like to acknowledge financial support of the European Union's Horizon 2020 research and innovation programme under the Marie Skłodowska-Curie grant agreement No 801604.

References

1. H.P. Garg, S.C. Mullick, V.K. Bhargava, *Solar Thermal Energy Storage*, Springer Science & Business Media (2012)
2. A. Sharma, V.V. Tyagi, C.R. Chen, D. Buddhi, Review on thermal energy storage with phase change materials and applications. Renew. Sustain. Energy Rev. **13**, 318–345 (2009)
3. B. Zalba, J.M. Marin, L.F. Cabeza, H. Mehling, Review on thermal energy storage with phase change: materials, heat transfer analysis and applications. Appl. Therm. Eng. **23**, 251–283 (2003)
4. S. Jegadheeswaran, S.D. Pohekar, Performance enhancement in latent heat thermal storage system: a review. Renew. Sustain. Energy Rev. **13**, 2225–2244 (2009)
5. X. Wang, Q. Guo, Y. Zhong, X. Wei, L. Liu, Heat transfer enhancement of neopentyl glycol using compressed expanded natural graphite for thermal energy storage. Renew. Energy. **51**, 241–246 (2013)
6. J. Sarwar, B. Mansoor, Characterization of thermophysical properties of phase change materials for non-membrane based indirect solar desalination application. Energy Convers. Manag. **120**, 247–256 (2016)
7. H.M. Ali, Experimental investigation on paraffin wax integrated with copper foam based heat sinks for electronic components thermal cooling. Int. Commun. Heat Mass Transf. **98**, 155–162 (2018)
8. J.-F. Su, X.-Y. Wang, S.-B. Wang, Y.-H. Zhao, Z. Huang, Fabrication and properties of microencapsulated-paraffin/gypsum-matrix building materials for thermal energy storage. Energy Convers. Manag. **55**, 101–107 (2012)
9. M. Pomianowski, P. Heiselberg, Y. Zhang, Review of thermal energy storage technologies based on PCM application in buildings. Energy Build. **67**, 56–69 (2013)
10. X. Tong, J.A. Khan, M. RuhulAmin, Enhancement of heat transfer by inserting a metal matrix into a phase change material. Numer. Heat Transf. Part A Appl. **30**, 125–141 (1996)
11. A. Mustaffar, D. Reay, A. Harvey, The melting of salt hydrate phase change material in an irregular metal foam for the application of traction transient cooling. Therm. Sci. Eng. Prog. **5**, 454–465 (2018)
12. D.C. Hyun, N.S. Levinson, U. Jeong, Y. Xia, Emerging applications of phase-change materials (PCMs): teaching an old dog new tricks. Angew. Chemie Int. Ed. **53**, 3780–3795 (2014)
13. A. Jamekhorshid, S.M. Sadrameli, M. Farid, A review of microencapsulation methods of phase change materials (PCMs) as a thermal energy storage (TES) medium. Renew. Sustain. Energy Rev. **31**, 531–542 (2014)
14. T. Uemura, N. Yanai, S. Watanabe, H. Tanaka, R. Numaguchi, M.T. Miyahara, Y. Ohta, M. Nagaoka, S. Kitagawa, Unveiling thermal transitions of polymers in subnanometre pores. Nat. Commun. **1**, 1–8 (2010)
15. C. Liu, Z. Rao, J. Zhao, Y. Huo, Y. Li, Review on nanoencapsulated phase change materials: preparation, characterization and heat transfer enhancement. Nano Energy. **13**, 814–826 (2015)
16. F. He, X. Wang, D. Wu, New approach for sol–gel synthesis of microencapsulated n-octadecane phase change material with silica wall using sodium silicate precursor. Energy **67**, 223–233 (2014)
17. X. Qiu, G. Song, X. Chu, X. Li, G. Tang, Preparation, thermal properties and thermal reliabilities of microencapsulated n-octadecane with acrylic-based polymer shells for thermal energy storage. Thermochim. Acta **551**, 136–144 (2013)
18. K. Tumirah, M.Z. Hussein, Z. Zulkarnain, R. Rafeadah, Nano-encapsulated organic phase change material based on copolymer nanocomposites for thermal energy storage. Energy **66**, 881–890 (2014)
19. F. Yu, Z.-H. Chen, X.-R. Zeng, Preparation, characterization, and thermal properties of microPCMs containing n-dodecanol by using different types of styrene-maleic anhydride as emulsifier. Colloid Polym. Sci. **287**, 549–560 (2009)

20. A. Sarı, C. Alkan, A. Biçer, A. Altuntaş, C. Bilgin, Micro/nanoencapsulated n-nonadecane with poly (methyl methacrylate) shell for thermal energy storage. Energy Convers. Manag. **86**, 614–621 (2014)
21. J. Shi, X. Wu, R. Sun, B. Ban, J. Li, J. Chen, Nano-encapsulated phase change materials prepared by one-step interfacial polymerization for thermal energy storage. Mater. Chem. Phys. **231**, 244–251 (2019)
22. M. Fuensanta, U. Paiphansiri, M.D. Romero-Sánchez, C. Guillem, Á.M. López-Buendía, K. Landfester, Thermal properties of a novel nanoencapsulated phase change material for thermal energy storage. Thermochim. Acta **565**, 95–101 (2013)
23. C. Chen, Z. Chen, X. Zeng, X. Fang, Z. Zhang, Fabrication and characterization of nanocapsules containing n-dodecanol by miniemulsion polymerization using interfacial redox initiation. Colloid Polym. Sci. **290**, 307–314 (2012)
24. A. Arshad, M. Jabbal, Y. Yan, J. Darkwa, The micro-/nano-PCMs for thermal energy storage systems: A state of art review. Int. J. Energy Res. **43**, 5572–5620 (2019)
25. Y. Qu, S. Wang, D. Zhou, Y. Tian, Experimental study on thermal conductivity of paraffin-based shape-stabilized phase change material with hybrid carbon nano-additives. Renew. Energy. **146**, 2637–2645 (2020)
26. D.D.W. Rufuss, L. Suganthi, S. Iniyan, P.A. Davies, Effects of nanoparticle-enhanced phase change material (NPCM) on solar still productivity. J. Clean. Prod. **192**, 9–29 (2018)
27. S. Sharma, L. Micheli, W. Chang, A.A. Tahir, K.S. Reddy, T.K. Mallick, Nano-enhanced Phase Change Material for thermal management of BICPV. Appl. Energy **208**, 719–733 (2017)
28. M. Nourani, N. Hamdami, J. Keramat, A. Moheb, M. Shahedi, Thermal behavior of paraffin-nano-Al2O3 stabilized by sodium stearoyl lactylate as a stable phase change material with high thermal conductivity. Renew. Energy. **88**, 474–482 (2016)
29. M. Li, A nano-graphite/paraffin phase change material with high thermal conductivity. Appl. Energy **106**, 25–30 (2013)

Chapter 4
Exploring the Relationship Between Heat Absorption and Material Thermal Parameters for Thermal Energy Storage

Law Torres Sevilla and Jovana Radulovic

Abstract Using thermal energy storage alongside renewables is a way of diminishing the energy lack that exists when renewable energies are unable to run. An in-depth understanding of the specific effect of material properties is needed to enhance the performance of thermal energy storage systems. In this paper, we used fitting models and regression analysis to quantify the effect that latent heat of melting and material density have on the overall heat absorption. A single tank system, with encapsulated phase change materials is analysed with materials properties tested in the range of values commonly found in the literature. These materials are, therefore, hypothetically constructed ones based on materials such as paraffin. The software used for the numerical analysis is COMSOL Mulitphysics. Results show that the relationship between the latent heat and density regarding heat absorbed is a positive linear function for this system.

Keywords Thermal storage · Energy · Materials · Graph fitting

4.1 Introduction

Every day the world is facing energy challenges whilst we further develop the use and implementation of renewables. Although these green technologies help stray away from fossil fuel usage, there are some major drawbacks. One is the intermittency issue, as they are only as effective as the weather allows them to be. However, combining thermal energy storage (TES) alongside them is a way of diminishing the mismatch between supply and demand.

Thermal energy storage systems can be divided into sensible heat, latent heat and thermochemical. This paper focuses on the combination of the first two. Sensible heat stores heat per degree increased, whereas latent heat stores heat during a phase change [1]. Sensible heat is the most used storage and the material selection is based on properties such as specific heat capacity, density and thermal conductivity. Latent heat

L. T. Sevilla (✉) · J. Radulovic
School of Mechanical and Design Engineering, University of Portsmouth, Anglesea Building, Anglesea Road, Portsmouth PO1 3DJ, UK
e-mail: law.torressevilla@port.ac.uk

© The Author(s) 2021
I. Mporas et al. (eds.), *Energy and Sustainable Futures*, Springer Proceedings in Energy,
https://doi.org/10.1007/978-3-030-63916-7_4

offers an advantage over sensible heat in terms of narrower operating temperatures and high energy density, requiring a material selection based on properties such as melting temperature and latent heat [2]. The most common type of latent heat is solid-liquid, hence why the analysis is on this particular one [3]. This paper focuses on parameters relevant to the heat absorption output, such as density (for sensible heat) and latent heat of melting (for latent heat). Designs for these are generally packed beds, single/double tank systems and heat exchangers. Common materials as storage mediums for sensible heat include solids such as rocks, ceramics, concrete, or fluids such as water, oils and inorganic salts [4]. For latent heat, materials used are classified into organic (paraffins, fatty acids and eutectics) and inorganic (salt hydrates and eutectics) [5]. Phase changing materials (PCMs) should have a number of desirable properties, amongst others: high thermal conductivity to enhance the heat transfer; they should be chemically stable, non-toxic or flammable; low cost and high availability are naturally advantageous [6]. Latent heat of melting and density are of vital importance for the performance of a TES. It is known that higher values lead to improved system performance. The aim of this study is to quantify the relationship between these thermal parameters and the heat absorption of a single tank system. This information can help in real life applications to aid users in material selection, material enhancement and contribute towards optimization of TES systems.

Parhizi et al. [7], similarly to the aim of this paper, assess the impact of thermal properties on a PCM based TES system for two models: a simplified cartesian 1D one and a cylindrical 3D one. They propose a theoretical heat transfer model, which aims to predict the rate of energy stored and energy density as functions of the thermal conductivity. Their results indicate that there is bound to be a trade-off when selecting the material. They report that while increasing thermal conductivity improves the rate of energy stored, the energy storage density itself does not change for the cartesian system and decreases for the cylindrical system.

4.2 Methodology

4.2.1 System Design

The analysed system consists of a symmetrical single cylindrical tank, packed with encapsulated and spherical PCMs. The tank height and length are 0.5 m and packed with a set of 19 × 17 encapsulated spheres containing the selected PCM. The tank frame is 0.025 m thick and the capsule is considered thin and negligible, with the sphere radius of 0.0125 m. It has a centric single inlet and outlet of dimensions 0.12 m and is symmetrical. The HTF is water and enters the system at a constant temperature of 90 °C and at a velocity of 0.01 m/s. The tank initially has still water inside at an ambient temperature of 20 °C. The analysis focuses on the centre point of the system, at coordinates (0,0) where the centre sphere is. The analysis was carried

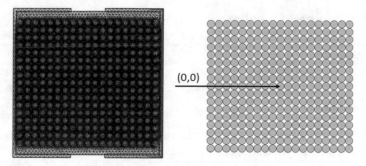

Fig. 4.1 System design meshed in COMSOL Multiphysics and coordinates (0,0)

out in COMSOL Multiphysics and the system was approximately 140,000 elements, mostly triangular prisms (Fig. 4.1).

4.2.2 Boundary Conditions and Assumptions

The 2D simulation is run in COMSOL Multiphysics for 60 min in 1 min step intervals. The model uses the "Laminar Flow" and "Heat Transfer in Fluids" physics, alongside the "Nonisothermal Flow" multi-physics. The numerical problem was solved using Fourier's Law and the heat equation for non-uniform isotropic mediums. Heat absorbed will then be calculated using Eq. 4.3.

$$\rho C_p \frac{\partial T}{\partial t} + \rho C_p u \cdot \nabla T + \nabla \cdot q = Q + Q_p + Q_{vd} \qquad (4.1)$$

$$q = k \nabla T \qquad (4.2)$$

$$q = m(Cps \cdot (Tm - Ti) + L + (Tf - Tm) \cdot Cpl) \qquad (4.3)$$

where ρ is density, Cp is heat capacity at constant pressure (subscripts "s" and "l" indicate solid and liquid phases), T is temperature (subscripts "m", "i" and "f" indicate melting, initial and final), t is time, u is velocity, q is heat flux, Q is the heat source, Qp is heat pressure work, Qvd is heat viscous dissipation and k is thermal conductivity.

The inlet boundary layer is velocity and the outlet is pressure, where initial pressure is zero and the model supress backflow. The wall boundary conditions are no slip and the tangential velocity is zero. The HTF is modelled to be laminar and incompressible and all materials are homogeneous and isotropic. Furthermore, there

Table 4.1 Material properties and their ranges for the tested hypothetical analysis

	Base Case	Variation 1	Variation 2	Variation 3	Variation 4
Melting temperature	45 °C	45 °C	45 °C	45 °C	45 °C
Latent Heat	200 kJ/kg	10 kJ/kg	500 kJ/kg	200 kJ/kg	200 kJ/kg
Density	800 kg/m³	800 kg/m³	800 kg/m³	600 kg/m³	1000 kg/m³

are no heat losses due to radiation and the outer wall of the tank is perfectly insulated. Lastly, the encapsulated spheres are modelled to be circles that do not undergo deformation.

4.2.3 Materials and Analysis

The materials used in this experiment are non-existing and hypothetical. They were constructed using informed, reasonable and educated values from materials found in systems and found in the literature. The base case is based on standard paraffin properties. In all simulations the melting temperature (45 °C), thermal conductivity (0.4 W/mK solid and 0.2 W/mK liquid), and specific heat capacity (2000 J/kgK solid and 2200 J/kgK liquid) were the same, yet density and latent heat were varied as shown in Table 4.1.

4.3 Results

Using the temperatures recorded for the centre sphere, the heat absorbed was calculated and a regression analysis was performed. Results shown in Figs. 4.2 and 4.3 indicate a linear relationship in both cases.

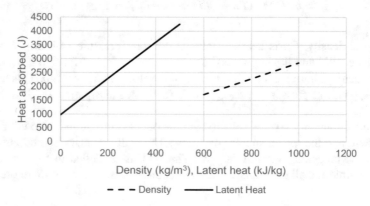

Fig. 4.2 Line plot of the heat absorbed (after 60 min) against latent heat and density

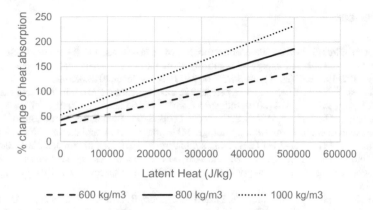

Fig. 4.3 Percentage change of the heat absorbed compared to the base case

4.4   Discussion

Both high density and latent heat yield higher heat absorption values.

The change in latent heat shows a much wider range in values, where the maximum and minimum are 4247.39 J and 1040.65 J (a difference of 3206.74 J), respectively. For density, the range was from 2854.88 J to 1713.14 J (a difference of 1141.74 J). This shows an increase from the minimum value of 308% for latent heat and 66% for density, approximately. Across the tested range, a relative increase of heat absorbed per 100 kJ/kg of latent heat was 6.5 kJ, compared to 2.9 kJ of heat absorbed per 100 kg/m^3 of density increase, showing that for low temperature heat storage the latent heat is a more influential factor. By increasing latent heat by 10% from the base case, the relative increase in heat absorption is 28%. In comparison, 24% increase is seen by density rise to 1000 kg/m^3 from 800 kg/m^3.

Paraffins and organic PCMs have desirable density values. However, high latent heat values considered in this study are outside the conventional range. While the addition of nanoparticles is a promising method for an alternation of the latent heat of PCMs [8], in practical systems it is beneficial to use high density PCMs to maximise heat absorption. While this increases the weight of the tank, if the volume variation during phase change is relatively low, the tank volume does not present a challenge.

4.5   Conclusion

In conclusion, increasing the latent heat and density without altering any other system parameters gives a simple linear relationship regarding heat absorption. Higher latent heat is a more influential parameter, although practically the same effect can be achieved with a material with higher density.

References

1. J.A. Almendros-Ibáñez, M. Fernández-Torrijos, M. Díaz-Heras, J.F. Belmonte, C. Sobrino, A review of solar thermal energy storage in beds of particles: packed and fluidized beds. Sol. Energy **192**, 193–237 (2019). https://doi.org/10.1016/j.solener.2018.05.047
2. E. Guelpa, V. Verda, Thermal energy storage in district heating and cooling systems: a review. Appl. Energy **252**, 113474 (2019). https://doi.org/10.1016/j.apenergy.2019.113474
3. A. Fallahi, G. Guldentops, M. Tao, S. Granados-Focil, S. Van Dessel, Review on solid-solid phase change materials for thermal energy storage: Molecular structure and thermal properties. Appl. Therm. Eng. **127**, 1427–1441 (2017). https://doi.org/10.1016/j.applthermaleng.2017.08.161
4. Q. Zhou, D. Du, C. Lu, Q. He, W. Liu, A review of thermal energy storage in compressed air energy storage system. Energy **188**, 115993 (2019). https://doi.org/10.1016/j.energy.2019.115993
5. A. Gil, M. Medrano, I. Martorell, A. Lazaro, P. Dolado, B. Zalba, L.F. Cabeza, State of the art on high temperature thermal energy storage for power generation. Part 1—concepts, materials and modellization. Renew. Sustain. Energy Rev. **14**, 31–55 (2010). https://doi.org/10.1016/j.rser.2009.07.035
6. I. Sarbu, C. Serbarchievici, A comprehensive review of thermal energy storage. Sustainability **10**, 191 (2018). https://doi.org/10.3390/su10010191
7. M. Parhizi, A. Jain, The impact of thermal properties on performance of phase change based energy storage systems. Appl. Therm. Eng. **162**, 114154 (2019). https://doi.org/10.1016/j.applthermaleng.2019.114154
8. A. Sari, C. Alkan, A.N. Ozcan, Synthesis and characterization of micro/nano capsules of PMMA/capric–stearic acid eutectic mixture for low temperature-thermal energy storage in buildings. Energy Buildings **90**, 106–113 (2015). https://doi.org/10.1016/j.enbuild.2015.01.013

Part II
ICT and Control

Chapter 5
A Novel Approach for U-Value Estimation of Buildings' Multi-layer Walls Using Infrared Thermography and Artificial Intelligence

Arijit Sen and Amin Al-Habaibeh

Abstract Estimating the U-value of walls of buildings is a key process to evaluate the overall thermal performance. Low U-value in buildings is desired in order to keep heat within the envelop and consume less energy in heating. Addressing the limitations in the currently used U-value estimation techniques, this paper proposes a novel approach for estimating the U-value of the envelop of buildings using infrared thermography and Artificial Neural Network (ANN) with the application of a point heat source. The novel system is calibrated by training the ANN in a lab environment using a wide range of samples with multi-layers to be able to estimate the in situ U-value of walls in real buildings during field work with relatively high accuracy.

Keywords U-value · Artificial intelligence · Neural network · Infrared thermography · Insulation

5.1 Introduction

The thermal performance of buildings is largely dependent on the thermal transmittance or U-value of their walls. The theoretical U-value of a building's wall is calculated as the reciprocal of the summation of thermal resistance values of different layers of the wall, where thermal resistance is a function of thermal conductivity and thickness of the layers [1].

$$U = \frac{1}{R_i + \frac{d_1}{k_1} + \frac{d_2}{k_2} + \cdots + R_e} \qquad (5.1)$$

Equation (5.1) represents the theoretical U value of the wall where, k is the thermal conductivity of the materials in different layers of the wall, d is the thickness of the

A. Sen (✉) · A. Al-Habaibeh
Product Innovation Centre (PIC), Nottingham Trent University, Nottingham, UK
e-mail: arijit.sen2016@my.ntu.ac.uk

A. Al-Habaibeh
e-mail: Amin.Al-Habaibeh@ntu.ac.uk

© The Author(s) 2021
I. Mporas et al. (eds.), *Energy and Sustainable Futures*, Springer Proceedings in Energy,
https://doi.org/10.1007/978-3-030-63916-7_5

wall's layers. R_i and R_e are the thermal resistance of air at internal surface and external surface respectively. The U-value of an existing building's wall often differs from its theoretical U-value because of the degradation of the wall materials, deviation of the layers' thicknesses from their designed values, presence of voids between two layers resulting from poor craftmanship and so on [2]. Also, in some countries the quality of materials used and type are not well-characterised in terms of thermal performance and consistency. There are different approaches exist for determining the U-value of existing buildings; however, they have some limitations [3]. One of the approaches, for example, requires collecting samples of wall materials for laboratory test by drilling holes through walls [3]. Two other non-invasive methods, namely: Heat Flux Meter (HFM) method and Infrared Thermovision Technique (ITT), are limited to be performed in winter only, as they require over 10°C temperature gradient between indoor and outdoor environment [4, 5]. Another approach of estimating U-value from external and internal wall temperatures is by using noncontact infrared sensors; this technique shows better performance over HFM method; however, it is restricted to use during night only and it requires the measurement to be conducted over a few days [6]. The reliable estimation of in situ U-value is difficult in real buildings because of many constraints such as, installing instruments, extended period of monitoring time, dependency on season, dependency on weather condition and presence of sunlight [4–6]. Furthermore, nonlinear and complex relationship among different parameters need to be considered for accurate estimation of U-value. Artificial Neural Network (ANN) could form a very strong tool to analyse nonlinear and complex relationship among input and output parameters and a previous research work shows successful use of infrared thermography and ANN with application of point heat for categorising walls with different U-values [3]. ANN can be trained to learn the relationship between inputs and outputs in a data set and apply the learning process to estimate the outputs from a similar unknown data set. The thermal profile that is generated from infrared images of walls with known U-value can be used to train ANN which later will be able to determine the U-value of an unknown wall. This paper presents a study with a novel product designed to estimate the U-value of a real building's wall from infrared images with application of point heat source using ANN, where the product is calibrated by training the ANN in laboratory environment with different wall samples.

5.2 Methodology

Wall sections with different U-values have been monitored with an infrared camera during the application of a point heat in the internal side of the walls. The thermal profiles generated from the monitored infrared images are used as the inputs for the ANN and theoretical U-values of those wall sections as the outputs to train ANN. Later, the thermal response of a real building's wall during the application of a similar point-heat in the internal side of the wall is monitored using infrared camera and the thermal profiles generated from the infrared image are analysed with the

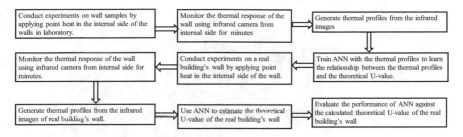

Fig. 5.1 Flow chart of the suggested methodology

ANN to estimate the U-value of the wall. The flow chart in Fig. 5.1 represents the methodology of this study. The ANN performance has been evaluated against the absolute deviation using Eq. (5.2).

$$\text{Absolute Deviation} = \frac{\left| U_{\text{predicted}} - U_{\text{calculated}} \right|}{U_{\text{calculated}}} \times 100\% \tag{5.2}$$

where $U_{\text{predicted}}$ is the ANN estimated U-value and $U_{\text{calculated}}$ is the theoretical U-value of real building's wall using Eq. (5.1). The training and evaluation of ANN have been performed 25 times to generalise the solution and the average value of them is considered as final outcome from the ANN.

5.3 Experimental Work

In this study six walls, indexed as A, B, C, D, E and F respectively, are monitored using infrared camera with the application of point heat source form internal side. Sample A is a concrete block wall without any internal or external mortar layers. Samples C and E are solid brick walls without any internal or external mortar layers. Samples B and D are concrete block and solid brick walls respectively with external insulation of 100 mm thick Ecotherm. Infrared images are captured using CHINO TP-L0260EN thermal camera at five seconds interval for about an hour resulting in around 720 images per set up of the experiment. Figure 5.2 shows the experimental work on one of the lab samples. The infrared camera is fitted into a plastic box which contains a diesel engine glow plug (Fig. 5.2). The glow plug acts as a point heat source and it is fitted within a plastic box in such a way that it always stays in touch with the internal wall surface. Two K-type thermocouples connected to NIUSB-TC01 data acquisition system are used to measure the glow plug's tip temperature and the ambient temperature respectively.

Experiments on samples A–E are conducted in lab environment and infrared images obtained from the experiments are used for the calibration process by training the ANN. The experiment on sample F is done inside a real building and infrared images captured during the experiment are used to estimate the U-value of the wall

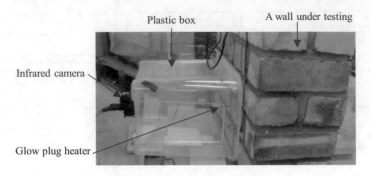

Fig. 5.2 The setup configuration of the experimental work

with the help of the calibrated ANN. The properties of the wall samples are listed in Table-1, where the U-values are calculated using Eq. (5.1). The values of R_i and R_e is taken as 0.13 and 0.04 respectively [7]. The thermal conductivity of concrete block in samples A and B are considered as 1.5 W/mK [8] and the thermal conductivity of the Ecotherm is 0.022 W/mK [9]. The thermal conductivity of brick in samples C and D are taken as 0.27 W/mK [10] and the thermal conductivity of the brick in samples E and F are taken as 0.56 for inner leave and 0.77 for outer leave W/mK [7] as the brick in the samples E and F are different than brick in the samples C and D.

5.4 Results and Discussion

Figure 5.3 represents the temperature profile and infrared image of samples D, E and F. It is noted in Fig. 5.3 that sample D has the warmest wall surface. As sample D has the lowest U-value, most of the heat is spread on the wall surface. Sample E has higher U-value than sample F and therefore, sample F is expected to have warmer wall surface. However, the infrared images in Fig. 5.3 convey the opposite information. This is also observed in the temperature profiles in Fig. 5.3.

This could happen due to the difference in ambient temperature during the time of experiments. Previous research work [3] also supports the fact and recommends three modified profiles as presented in Eqs. (5.3), (5.4) and (5.5).

$$T^a_{(i,j,k)} = T_{(i,j,k)} - T^{ext}_k \qquad (5.3)$$

$$T^b_{(i,j,k)} = \sum_1^k \left[T_{(i,j,k+1)} - T_{(i,j,k)} \right] \qquad (5.4)$$

$$T^{ab}_{(i,j,k)} = \sum_1^k \left[T^a_{(i,j,k+1)} - T^a_{(i,j,k)} \right] \qquad (5.5)$$

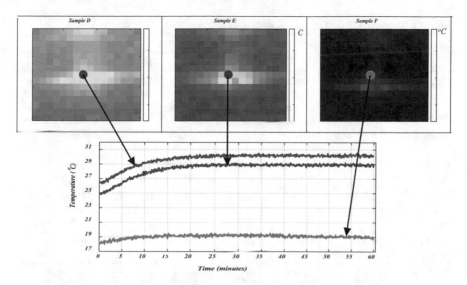

Fig. 5.3 A comparison between temperature profiles: sample D, E and F

Here $T_{(i,j,k)}$ is the original temperature value at pixel *(i,j,k)* on infrared image, $T^a_{(i,j,k)}$ is the modified temperature value of pixel *(i,j,k)* on the infrared image, $T^b_{(i,j,k)}$ is the cumulative temperature difference at pixel *(i,j,k)*, $T^{ab}_{(i,j,k)}$ is the cumulative temperature difference of profile T^a and T^{ext}_k is the external temperature at the time of capture of the corresponding infrared image. In Eqs. (5.3) to (5.5), *i* and *j* refer to the pixel indices of infrared image in x and y direction respectively, and *k* refers to the image index in the sequence of infrared images. A feed forward neural network with single hidden layer containing 20 neurons is developed using MATLAB. Sigmoid transfer function is used in the neurons of hidden layer and Levenberg-Marquardt backpropagation algorithm is used as the learning algorithm. The standard deviation of profile T, T^a T^b and T^{ab} are selected as the input parameters to the neural network and the U-value listed in Table 5.1 as the output. The training data set contains 60 samples of each profile taken equally from the walls AE mentioned in Table 5.1. The test data set is composed of 12 samples from each profile of the wall F. Figure 5.4 shows the average training performance of ANN for profile T, T^a, T^b and T^{ab} of samples A-E. The absolute deviation ranges between 5% to 15% with profile T having the lowest absolute deviation and profile T^{ab} having the highest.

Figure 5.5 represents the ANN's performance regarding predicting the U-value of sample F. The theoretical U-values are the target for the ANN and the ANN predicted U-values are the achievement made. The bar chart in Fig. 5.5a compares the performance of ANN with the target values. It is found in Fig. 5.5a, that the ANN produces the closest prediction to the theoretical U-value in profile T^b. Profile T^b and T^{ab} achieve the lowest absolute deviation which is around 20% (Fig. 5.5b). The training performance of ANN for profile T is found to be the best but the test performance is the worst among all profiles, which indicates that ANN overfits this

Table 5.1 Properties of wall samples used to train and test ANN

Sample Number	Training Samples					Test Sample
	A	B	C	D	E	F
Material	Concrete block wall	Concrete block + External insulation with Ecotherm	Brick	Brick + External insulation with Ecotherm	Solid Brick Wall	Inner mortar layer + Solid Brick Wall + Outer mortar layer
Thickness (mm)	95	95 + 100 = 195	100	100 + 100 = 200	230	15 + 230 + 20 = 265
Thermal Conductivity (W/mK)	1.5	1.5 & 0.22	0.27	0.27 & 0.22	0.56* & 0.77**	0.88,0.56, & 0.94 respectively.
U-value (W/m²K)	4.29	1.45	1.86	1.01	1.92	1.62
Cross section						

*Inner layer, **Outer layer

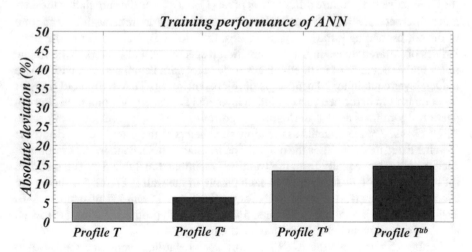

Fig. 5.4 Average training performance of ANN for different profiles of samples A-E

profile. On the other hand, the difference between the training performance and the test performance are found to be the lowest for profile T^{ab}, which indicates the best fit among all profiles. Similarly, for profile T^{b}, ANN shows a reasonable difference between training performance and test performance indicating a better fit

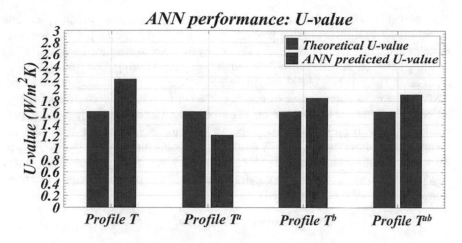

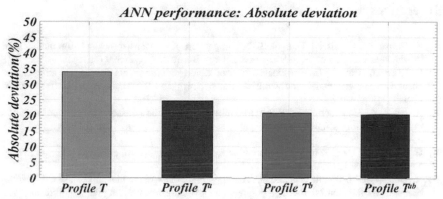

Fig. 5.5 **a** A comparison between theoretical and predicted U-values; **b** The absolute deviations of different profiles

than the two others. In profile T^a, the difference between training performance and test performance is significantly high indicating an overfit as well. The profiles T^{ab} and T^b are developed considering the cumulative temperature difference; and thus, it generalises the thermal behaviour of walls more precisely. As a result, the ANN produces the best performance for those two profiles.

5.5   Conclusion

The U-value of buildings' walls signifies the thermal performance and the energy transferred via the envelop. There is a need to estimate the U value of buildings' components in situ and in a rapid way. This paper presents a novel approach of determining U-value of a real building's wall by designing a novel product that can be calibrated using the training process of an ANN with the help from material samples in the lab. The results signify the ability of ANN to estimate U-value of real building based on the analysis of infrared data captured during the application of a point heat. There is no general rule to decide on the optimum configuration of an ANN or the most efficient size of training data set required to perform successful ANN analysis; as this depends on the nature of the utilised data as input and target values. The results presented in this paper show that ANN is capable of estimating U-value of buildings' wall with 80% accuracy, including multi-layered walls. Future work will be utilised to attempt to enhance the accuracy of the system using other ANN architectures, including more training data, and compare between the existing technologies and the this suggested one.

References

1. K. Gaspar, M. Casals, M. Gangolells, A comparison of standardized calculation methods for in situ measurements of façades U-value. Energy Build. **130**, 592–599 (2016)
2. L. Evangelisti et al., In situ thermal transmittance measurements for investigating differences between wall models and actual building performance. Sustainability **7**(8), 10388–10398 (2015)
3. A. Sen, A. Al-Habaibeh, The design of a novel approach for the assessment of thermal insulation in buildings using infrared thermography and artificial intelligence. Int. J. Design Eng. **9**(1), 65–77 (2019)
4. G. Ficco et al., U-value in situ measurement for energy diagnosis of existing buildings. Energy Build. **104**, 108–121 (2015)
5. R. Albatici, A.M. Tonelli, M. Chiogna, A comprehensive experimental approach for the validation of quantitative infrared thermography in the evaluation of building thermal transmittance. Appl. Energy **141**, 218–228 (2015)
6. N. Sakkas et al., Non intrusive U value metering. Open J. Energy Effic. **4**, 28–35 (2015)
7. B. Anderson, Conventions for U-value calculations 2006 Edition UK: BRE Press Scotland (2006)
8. ISO FDIS 10456 (2007) Building materials and products—Hygrothermal properties—Tabulated design values and procedures for determining declared and design thermal values. http://www.superhomes.org.uk/wp-content/uploads/2016/09/Hygrothermal-properties.pdf. Accessed May 14, 2018
9. EcoTherm Rigid Thermal Insulation Boards. https://www.ecotherm.co.uk/about-us/features-benefits. Accessed April 29, 2019
10. K.D. Antoniadis, M.J. Assael, C.A. Tsiglifisi, S.K. Mylona, Improving the design of greek hollow clay bricks. Int. J. Thermophys. **33**(12), 2274–2290 (2012)

Chapter 6
Binary versus Multiclass Deep Learning Modelling in Energy Disaggregation

Pascal A. Schirmer and Iosif Mporas

Abstract This paper compares two different deep-learning architectures for the use in energy disaggregation and Non-Intrusive Load Monitoring. Non-Intrusive Load Monitoring breaks down the aggregated energy consumption into individual appliance consumptions, thus detecting device operation. In detail, the "One versus All" approach, where one deep neural network per appliance is trained, and the "Multi-Output" approach, where the number of output nodes is equal to the number of appliances, are compared to each other. Evaluation is done on a state-of-the-art baseline system using standard performance measures and a set of publicly available datasets out of the REDD database.

Keywords Energy disaggregation · Non-Intrusive load monitoring · Deep-Learning · Deep neural networks

6.1 Introduction

Due to global warming average temperatures are rising and several techniques for energy reduction have been proposed in order to reduce the total energy consumption. However, to make use of those techniques accurate and fine-grained monitoring of electrical energy consumption is needed [1], since the energy consumption of most households is monitored via monthly aggregated measurements and thus cannot provide real-time feedback. Moreover, according to [2] the largest improvements in terms of energy savings can be made when monitoring energy consumption on device level. The term Non-Intrusive Load Monitoring (NILM) is used to describe the estimation of the power consumption of individual appliances, based on a single measurement on the inlet of a household or building [3]. In contrast to NILM, the term

P. A. Schirmer (✉) · I. Mporas
School of Engineering and Computer Science, University of Hertfordshire, Hatfield AL10 9AB, UK
e-mail: p.schirmer@herts.ac.uk

I. Mporas
e-mail: i.mporas@herts.ac.uk

© The Author(s) 2021
I. Mporas et al. (eds.), *Energy and Sustainable Futures*, Springer Proceedings in Energy,
https://doi.org/10.1007/978-3-030-63916-7_6

Intrusive Load Monitoring (ILM) is used when multiple sensors are used, usually one per device. ILM compared to NILM has the drawback of higher cost through wiring and data acquisition making it unsuitable for monitoring households where appliances can change. Conversely, NILM has the goal of finding the inverse of the aggregation function through a disaggregation algorithm using as input only the aggregated power consumption which makes it a highly underdetermined problem and thus impossible to solve analytically [4].

Several NILM methodologies based on deep neural networks have been proposed in the literature, e.g. Convolutional Neural Networks (CNNs) [5], Recurrent Neural Networks (RNNs) [6] and Long Short Time Memory (LSTM) [7]. Additionally, combinations of machine learning algorithms for fusion of information [8] and modelling of temporal dynamics [9] have also been proposed, especially for low sampling frequencies [10]. Particularly, these models operate either according to the "One versus All" approach, where one Deep Neural Network (DNN) per appliance is trained or the "Multi-Output" approach, where the number of output nodes is equal to the number of appliances. As it is not clear which architecture leads to better performances a comparison of these two architectures is needed.

The remainder of this paper is organized as follows: In Sect. 6.2 the two NILM systems based on DNNs are presented. In Sect. 6.3 the experimental setup is described and in Sect. 6.4 the evaluation results are presented. Finally, the paper is concluded in Sect. 6.5.

6.2   Proposed Architecture

NILM energy disaggregation can be formulated as the task of determining the power consumption on device level based on the measurements of one sensor, within a time window (frame or epoch). Specifically, for a set of $M - 1$ known devices each consuming power p_m, with $1 \leq m \leq M - 1$, the aggregated power p_{agg} measured by the sensor will be [11]:

$$p_{agg} = f(p_1, p_2, \ldots, p_{M-1}, g) = \sum_{m=1}^{M-1} p_m + g = \sum_{m=1}^{M} p_m \qquad (6.1)$$

where $g = p_M$ is a 'ghost' power consumption usually consumed by one or more unknown devices. In NILM the goal is to find estimations $\hat{p}_m$, $\hat{g}$ of the power consumption of each device m using an estimation method f^{-1} with minimal estimation error [11], i.e.

$$\hat{P} = \{\hat{p}_1, \hat{p}_2, \ldots, \hat{p}_{M-1}, \hat{g}\} = f^{-1}(g(p_{agg}))$$

$$\text{s.t. argmin}_{f^{-1}} \left\{ (p_{agg} - \sum_1^M \hat{p}_m)^2 \right\} \qquad (6.2)$$

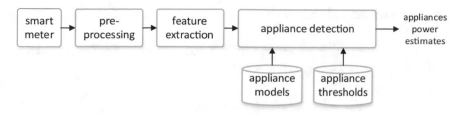

Fig. 6.1 Block diagram of the NILM architecture, including the smart meter (data source), pre-processing, feature extraction and appliance detection

where p_{agg} is the aggregated power consumption, p_m the power consumption of m-th device with $p_g = p_M$ being the 'ghost' power consumption, $\hat{P} = \{\hat{P}_m, \hat{P}_g\}$ the estimates of the per device power consumptions, f^{-1} an estimation method and $g()$ a function transforming a time window of the aggregated power consumption into a multidimensional feature vector $F \in \mathbb{R}^N$. The block diagram of the NILM architecture adopted in the present evaluation is illustrated in Fig. 6.1 and consists of three stages, namely the pre-processing, feature extraction and appliance detection.

In detail, the aggregated power consumption signal calculated from a smart meter is initially pre-processed i.e. passed through a median filter [12] and then frame blocked in time frames. After pre-processing feature vectors, F of dimensionality N, one for each frame are calculated. In the appliance detection stage, the feature vectors are processed by a regression algorithm using a set of pre-trained appliance models to estimate the power consumption of each device. The output of the regression algorithm estimates the corresponding device consumption and a set of thresholds, T_m with $1 \leq m \leq M$ with $T_g = T_M$, for each device including the ghost device ($m = M$) is used to decide whether a device is switched on or off. In the present evaluation the estimation method is implemented using two different deep-learning architectures as shown in Fig. 6.2.

As can be seen in Fig. 6.2 the two architectures only differ in their number of output nodes with architecture (a) using a single output node and one DNN per device and architecture (b) using one output node per device and a single DNN for all devices.

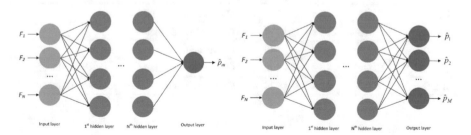

Fig. 6.2 Two deep neural network architectures using **a** one node and **b** M nodes in the output layer

6.3 Experimental Setup

The NILM architecture presented in Sect. 6.2 was evaluated using several publicly available datasets and a deep neural network for regression.

a. Datasets

To evaluate performance five different datasets of the REDD [13] database were used. The REDD database was chosen as it contains power consumption measurements per device as well as the aggregated consumption. The REDD-5 dataset was excluded as its measurement duration is significantly shorter than the rest of the datasets in the REDD database [14]. The evaluated datasets and their characteristics are tabulated in Table 6.1 with the number of appliances denoted in the column #App. In the same column, the number of appliances in brackets is the number of appliances after excluding devices with power consumption below 25 W, which were added to the power of the ghost device, similarly to the experimental setup followed in [15]. The next three columns in Table 6.1 are listing the sampling period T_s, the duration T of the aggregated signal used and the appliance type for each evaluated dataset. The appliances type categorization is based on their operation as described in [11].

b. Pre-processing and Parameterization

During pre-processing the aggregated signal was processed by a median filter of 5 samples as proposed in [12] and then was frame blocked in frames of 10 samples with overlap between successive frames equal to 50% (i.e. 5 samples). Specifically, raw samples have been used at the input stage of the DNN, thus $F_{1,...,N}$ being the raw samples of each frame respectively. Furthermore, the number of hidden layers for each architecture was optimized using a bootstrap training dataset resulting into an architecture with 3 hidden layers and 32 sigmoid nodes for (a) and 2 hidden layers and 32 sigmoid nodes for (b).

Table 6.1 List of the evaluated datasets and their corresponding properties

Dataset	Properties			
	# App.	T_s	T	Appliance type
REDD-1	18 (17)	3 s	7d	One state/multi state/non-linear
REDD-2	9 (10)	3 s	11d	One state/multi state
REDD-3	20 (18)	3 s	14d	One state/multi state/non-linear
REDD-4	18 (16)	3 s	14d	One state/multi state/continuous/non-linear
REDD-6	15 (14)	3 s	12d	One state/multi state/continuous/non-linear

Table 6.2 Performance evaluation of 5 datasets for two deep learning architectures using estimation accuracy

Dataset	REDD-1 (%)	REDD-2 (%)	REDD-3 (%)	REDD-4 (%)	REDD-6 (%)	AVG (%)
One versus All	71.38	**86.00**	70.44	**68.21**	**87.17**	**76.64**
Multi-output	**71.87**	85.65	**71.69**	67.18	85.03	76.28

6.4   Experimental Results

The NILM architecture presented in Sect. 6.2 was evaluated according to the experimental setup described in Sect. 6.3. The performance was evaluated in terms of appliance power estimation accuracy (E_{ACC}), as proposed in [13] and defined in Eq. 6.3. The accuracy estimation is considering the estimated power $\hat{p}_m$ for each device m, where T is the number of frames and M is the number of disaggregated devices.

$$E_{ACC} = 1 - \frac{\sum_{t=1}^{T} \sum_{m=1}^{M} \left| \hat{p}_m^t - p_m^t \right|}{2 \sum_{t=1}^{T} \sum_{m=1}^{M} \left| p_m^t \right|} \tag{6.3}$$

To compare the two architectures the publicly available REDD database is used [13]. The results are tabulated in Table 6.2.

As can be seen in Table 6.2 the performance of datasets with smaller number of appliances (e.g. REDD-2/6) is significantly higher than for the datasets with higher number of appliances (e.g. REDD-1/3/4). Furthermore, the "One versus All" approach slightly outperforms the "Multi-Output" approach performing 0.36% better on average. However, it has to be mentioned that the "One versus All" approach requires the training of M deep neural networks resulting into significantly higher training times.

6.5   Conclusion

In this paper two different deep learning architectures for non-intrusive load monitoring were compared. Specifically, the "One versus All" approach using one deep neural network per appliance was compared to the "Multi-Output" approach using one deep neural network with the same number of output nodes than the number of appliances. It was shown, that both architectures have similar performance with average accuracies of 76.6% for the "One versus All" approach and 76.3% for the "Multi-Output" approach respectively. However, in terms of training time it must be considered, that for the "One versus All" approach M deep neural networks must be trained, resulting in significant higher training times.

Acknowledgements This work was supported by the UA Doctoral Training Alliance (https://www. unialliance.ac.uk/) for Energy in the United Kingdom.

References

1. A. Chis, J. Rajasekharan, J. Lunden, V. Koivunen, Demand response for renewable energy integration and load balancing in smart grid communities, in *2016 24th European Signal Processing Conference (EUSIPCO)*, (2016), pp. 1423–1427
2. M.N. Meziane, K. Abed-Meraim, Modeling and estimation of transient current signals, in *2015 23rd European Signal Processing Conference (EUSIPCO)*, (2015), pp. 1960–1964
3. G.W. Hart, Nonintrusive appliance load monitoring. Proc. IEEE **80**(12), 1870–1891 (1992). https://doi.org/10.1109/5.192069
4. M. Gaur, A. Majumdar, Disaggregating transform learning for non-intrusive load monitoring. IEEE Access **6**, 46256–46265 (2018). https://doi.org/10.1109/ACCESS.2018.2850707
5. A. Harell, S. Makonin, I.V. Bajic, Wavenilm: a causal neural network for power disaggregation from the complex power signal, in *ICASSP 2019 - 2019 IEEE International Conference on Acoustics, Speech and Signal Processing (ICASSP)*, (2019), pp. 8335–8339
6. P.A. Schirmer, I. Mporas, Energy Disaggregation Using Fractional Calculus, in *ICASSP 2020 - 2020 IEEE International Conference on Acoustics, Speech and Signal Processing (ICASSP)* (Barcelona, Spain, 2020), Apr. 2020–Aug. 2020, pp. 3257–3261
7. M. Kaselimi, N. Doulamis, A. Doulamis, A. Voulodimos, E. Protopapadakis, Bayesian-optimized Bidirectional LSTM regression model for Non-intrusive Load monitoring, in *ICASSP 2019 - 2019 IEEE International Conference on Acoustics, Speech and Signal Processing (ICASSP)*, (2019), pp. 2747–2751
8. P.A. Schirmer, I. Mporas, A. Sheikh-Akbari, Energy disaggregation using two-stage fusion of binary device detectors. Energies **13**(9), 2148 (2020). https://doi.org/10.3390/en13092148
9. P.A. Schirmer, I. Mporas, A. Sheikh-Akbari, Robust energy disaggregation using appliance-specific temporal contextual information. EURASIP J. Adv. Signal Process. **2020**(1), 394 (2020). https://doi.org/10.1186/s13634-020-0664-y
10. P.A. Schirmer, I. Mporas, energy disaggregation from low sampling frequency measurements using multi-layer zero crossing rate, in *ICASSP 2020 - 2020 IEEE International Conference on Acoustics, Speech and Signal Processing (ICASSP)*, (Barcelona, Spain, 2020) Apr. 2020–Aug. 2020, pp. 3777–3781
11. P.A. Schirmer, I. Mporas, Statistical and electrical features evaluation for electrical appliances energy disaggregation. Sustainability **11**(11), 3222 (2019). https://doi.org/10.3390/su11113222
12. C. Beckel, W. Kleiminger, R. Cicchetti, T. Staake, S. Santini, The ECO data set and the performance of non-intrusive load monitoring algorithms, in *BuildSys'14: Proceedings of the 1st ACM Conference on Embedded Systems for Energy-Efficient Buildings*, (2014), pp. 80–89
13. J.Z. Kolter, M.J. Johnson, eds., *REDD: A Public Data Set for Energy Disaggregation Research*, (2011)
14. V. Andrean, X.-H. Zhao, D.F. Teshome, T.-D. Huang, K.-L. Lian, A hybrid method of cascade-filtering and committee decision mechanism for non-intrusive load monitoring. IEEE Access **6**, 41212–41223 (2018). https://doi.org/10.1109/ACCESS.2018.2856278
15. S.M. Tabatabaei, S. Dick, W. Xu, Toward non-intrusive load monitoring via multi-label classification. IEEE Trans. Smart Grid **8**(1), 26–40 (2017). https://doi.org/10.1109/TSG.2016.2584581

Chapter 7
Review of Heat Demand Time Series Generation for Energy System Modelling

Malcolm Peacock, Aikaterini Fragaki, and Bogdan J. Matuszewski

Abstract National heat demand time series are important inputs into national energy system models. Although time series for primary fuel such as gas might be available, heat demand is not and measuring heat demand is only possible for individual buildings. Four different methods are used in this work to generate daily heat demand time series for Great Britain for 2016–2018 from temperature and windspeed and are validated against heat demand derived from national grid gas demand. All seem to model heat demand well.

Keywords Heat · Demand · Time series · Energy system · Modelling

7.1 Introduction

Energy system models [1, 2] used to investigate renewable electricity and heat at national level require knowledge of heat demand. Whilst estimating the heat demand of one building is possible by measuring internal and external temperatures and the input fuel energy, knowing the heat demand of an entire country is very difficult [3]. Bottom up statistical models using regression from measured data have been used to generate heat demand time series for periods up to 12 months [3–5] but tend to be limited to the year of the measurements. Bottom up aggregated thermal models have uncertainty over the many different parameters that need to be specified [6] and have difficulty capturing diversity on a national scale [7]. Multi-year daily national heat demand time series are typically generated top down [4] using methods which have few inputs apart from weather parameters and annual fuel demand. Gas energy time series are used for validation.

Top down methods use national heat demand from a reference year derived from fuel sales figures. Annual fuel sales are divided into end use based on a combination of consumer surveys, building measurements and modelling. The annual demand is split up into days using historic weather data [4]. Standard hourly heat demand profiles

M. Peacock (✉) · A. Fragaki · B. J. Matuszewski
School of Engineering, University of Central Lancashire, Preston PR12HE, UK
e-mail: MPeacock2@uclan.ac.uk

© The Author(s) 2021   53
I. Mporas et al. (eds.), *Energy and Sustainable Futures*, Springer Proceedings in Energy,
https://doi.org/10.1007/978-3-030-63916-7_7

Table 7.1 2018 annual fuel
use TWh

Fuel	Gas	Other	Total
Domestic space	191	68	259
Services space	56	36	92
Domestic water	56	12	82
Services water	7	6	15
Non heat use	98		
Total	408		

based on observations and modelling are then applied to each day to generate an hourly time series. This paper compares four typical top down methods of generating multi-year daily heat demand by generating time series for Great Britain.

7.2 Input Data Used in This Work

Monthly mean wind speeds for 1979 to 2018 and ambient air temperatures every 6 h for 2016–2018 were taken from the ERA 5 interim weather reanalysis [8] at a spatial resolution of 0.75° × 0.75° for Great Britain. The distribution of the Great Britain population on a 1 km grid is taken from Eurostat [9] for 2011.

UK annual fuel demand figures for space heating and hot water were obtained from [10] and multiplied by 0.99 to convert to values for Great Britain, the Northern Ireland gas usage [11] being only 1% of the UK. Northern Ireland has its own gas network, and therefore this study concentrates on Great Britain rather than the UK.

Table 7.1 shows these annual demand figures converted to heat demand assuming the following efficiencies: gas 80%, oil 85%, solid fuel 76%, electricity 100%, heat (e.g. combined heat and power) 100%, bioenergy and waste 87%.

The Non-Daily Metered (NDM) daily gas demand time series for 2016–2018 was taken from national grid gas data explorer [12] and excludes "most gas fired power stations and some large industrial units". It includes agriculture and some industrial space heating which may explain why the sum of this time series for 2018 of 435 TWh does not match the total gas energy from Table 7.1 of 408 TWh.

7.3 Top Down Methods

In top down methods annual space heating demand is usually split using a temperature dependent equation as in [13]. Water heating is either done in a similar way or sometimes just split equally between days [14]. Sometimes only a single daily temperature for the UK is used [14, 15], but a more sophisticated method is to use weather reanalysis data weighted by population density at weather grid points [13,

16]. To account for the thermal inertia of buildings an effective temperature including the temperature of previous days is often used [3, 13, 17].

Four typical top down methods were selected from a survey of UK energy system studies from the last 10 years. The selection was made to incorporate all the techniques used in multi-year UK studies.

- Regression equation based on building measurements from Watson et al. [3] **(Watson)**
- Gas demand methodology from German Association of Energy and Water Industries (BDEW) as used by Ruhnau [13] **(BDEW)**
- Heating degree days (HDD) with a base temperature of 15.5 as used by Barton et al. [15] and Staffell et al. [15, 16] **(HDD 15.5)**
- Heating degree days with a base temperature of 12.8 as used by Hooker-Stroud et al. [14] **(HDD 12.8)**

The python program used by Ruhnau [13] to generate heat demand time series using the BDEW method was modified so that it would generate heat demand time series for all four methods (modified code available at https://github.com/malcolmpe acock/heat).

A reference temperature (Eq. (7.1)) was calculated at each weather grid point (l) and day (d) based on the ambient temperatures of the N previous days to account for the thermal inertia of buildings (for $d < N$, $N = d$).

$$T_{d,l}^{Ref} = \frac{\sum_{n=0}^{N} 0.5^n T_{d-n,l}^{amb}}{\sum_{n=0}^{N} 0.5^n} \qquad (7.1)$$

where $T_{d,l}^{Ref}$ is the reference temperature for day d at location l and T^{amb} is the mean ambient air temperature for that location and day. The daily heat demand was calculated by summing up the demand values for all locations and weighting by population (mapped onto the weather grid), Eq. (7.2)

$$HDT_d = \frac{HD_{annual}}{P_{total} \cdot f_{total}} \sum_{l=0}^{NL} f_{d,l} \cdot P_l \qquad (7.2)$$

where $f_{d,l}$ the daily demand factor for day d and location l is derived differently for each method and for space and water heating as show in Table 7.2. HDT_d is the heat demand for day d, P_l is the population at location l, P_{total} is the total population, NL is the number of locations, HD_{annual} is the annual heat demand derived from Table 7.1, and f_{total} is the sum of all the demand factors.

In Table 7.2 T_0 is 40 °C and A, B, C, D, m_{space}, m_{water}, b_{space}, b_{water} are factors taken from the code download for [13]. These factors depend on (i) UK 40 year mean wind speed and (ii) type of building (domestic: multi-family house 30%/single family house 70% or commercial building).

Table 7.2 Temperature equations to factor annual heat demand

Method	Demand factor equation	Reference temperature
BDEW space [13]	$f_{d,l} = \dfrac{A}{1+\left\{\frac{B}{T_{d,l}^{Ref}-T_0}\right\}^c} + D +$ $max\begin{pmatrix} m_{space} - T_{d,l}^{Ref} + b_{space} \\ m_{water} - T_{d,l}^{Ref} + b_{water} \end{pmatrix}$	Current day and 3 previous days (N = 3)
BDEW water [13]	$f_{d,l} =$ $\begin{pmatrix} D + m_{water}.T_{d,l}^{ref} + b_{water} \ T_{d,l}^{ref} > 15°C \\ D + m_{water}.15 + b_{water} \ T_{d,l}^{ref} \le 15°C \end{pmatrix}$	(N = 3)
Watson space [3]	$f_{d,l} =$ $\begin{cases} -6.71 T_{d,l}^{Ref} + 111, \ for \ T_{d,l}^{Ref} < 14.1°C \\ -1.21 T_{d,l}^{Ref} + 33, \ for \ T_{d,l}^{Ref} > 14.1°C \end{cases}$	1 previous day (N = 1)
Watson water [3]	$f_{d,l} = -0.0458 T_{d,l}^{Ref} + 1.8248$	(N = 1)
HDD 15.5 space [16]	$f_{d,l} = \begin{cases} 15.5 - T_{d,l}^{Ref}, \ for \ T_{d,l}^{Ref} < 15.5°C \\ 0, \ for \ T_{d,l}^{Ref} > 15.5°C \end{cases}$	Current day only (N = 0)
HDD 15.5 water [16]	$f_{d,l} = 1.0$	
HDD 12.8 space [14]	$f_{d,l} = \begin{cases} 12.8 - T_{d,l}^{Ref}, \ for \ T_{d,l}^{Ref} < 12.8°C \\ 0, \ for \ T_{d,l}^{Ref} > 12.8°C \end{cases}$	Current day only (N = 0)
HDD 12.8 water [14]	$f_{d,l} = 1.0$	

Space and water heating are then aggregated to give a final heat demand time series. In order to make a comparison with a heat demand time series generated from gas, space heat demand was scaled by 0.72 because only 72% of space heating comes from gas and water heating by 0.81 (derived from Table 7.1)

7.4 Validation Using Heat Demand from Gas

A heat demand time series was generated from NDM gas demand using Eq. (7.3)

$$HDG_d = \left\{0.8G_d - \left(\frac{0.8G_T - G_H}{365}\right)\right\} \qquad (7.3)$$

where HDG_d is heat demand, G_d is the daily gas demand, 0.8 scales for boiler efficiency consistent with Table 7.1, G_T is the sum of the G_d values for the year, and G_H is the sum of the annual gas space and water heating from Table 7.1. This assumes that $0.8G_T - G_H$ is non-heat gas and is therefore not weather dependent and can be split equally between the 365 days of the year. The heat time series generated using the 4 methods were compared with this gas derived heat demand for 2016, 2017 and 2018

7.5 Results and Discussion

Heat demand time series generated from temperature are commonly validated by showing an $R^2 > 0.95$ correlation with time series derived from historic gas energy [3, 4, 13]. Table 7.3 shows R^2 calculated using the python statsmodels.OLS function [18] of 0.97 or greater for all 4 methods. This suggests for example, that 99% of the variation in gas energy is explained by the variation in heat demand for BDEW 2016.

The last two rows of Table 7.3 show experiments simplifying HDD 15.5. **HDD 15.5 1D** shows the results of using only the temperature of the current day (N = 0 in Eq. (7.1)). **HDD 15.5 1T** uses one mean temperature for the whole of Great Britain with no weighting by population (NL = 1 in Eq. (7.2)). This suggests that the additional effort of applying more complex methods may only give a small gain in accuracy.

The load duration curve, Fig. 7.1 plots the time series sorted by heat demand (instead of time) and suggests that the gas derived demand has lower troughs than the heat demand. The HDD 15.5 appears to match the gas series most closely and it also has the smallest RMSE for 2018.

Figure 7.2 shows all 4 methods compared against the expected heat demand derived from gas energy for 2018. There tends to be over prediction in summer and

Table 7.3 Comparison of methods

Method	Year	RMSE	R^2
BDEW	2016	0.18	0.99
Watson	2016	0.22	0.98
HDD 15.5	2016	0.17	0.99
HDD 12.8	2016	0.17	0.99
BDEW	2017	0.18	0.98
Watson	2017	0.22	0.97
HDD 15.5	2017	0.17	0.98
HDD 12.8	2017	0.17	0.97
BDEW	2018	0.16	0.98
Watson	2018	0.20	0.97
HDD 15.5	2018	0.16	0.98
HDD 12.8	2018	0.21	0.97
HDD 15.5 1D	2018	0.20	0.97
HDD 15.5 1T	2018	0.20	0.97

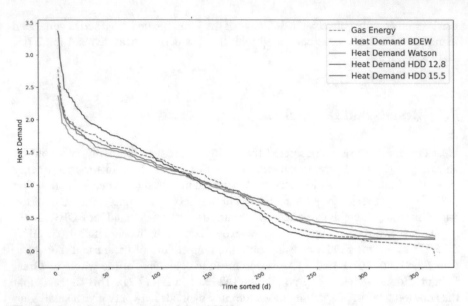

Fig. 7.1 Load duration curve heat demand 2018

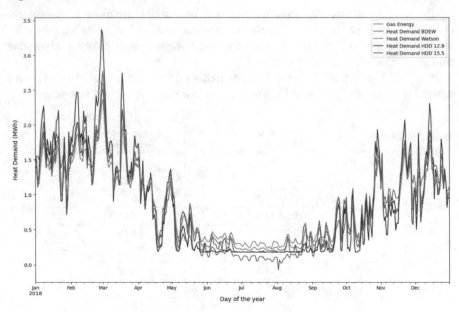

Fig. 7.2 Heat demand (HDT$_d$) and gas heat demand (HDG$_d$) daily time series for 2018

under prediction in winter. The low values between 2nd and 5th August suggest underestimation of the weather dependent part of the gas time series in Eq. (7.3).

7.6 Conclusions

Heat demand time series were generated from three years temperature and windspeed time series using four existing top down methods. All four methods seem to represent the observed data equally well. This suggests heat demand time series from any of these methods could be used as input to a national energy model. It was shown that simpler methods using a mean Great Britain temperature without thermal inertia also produce good results. However, considering the wider context of the present work, it will be the implementation of these generated heat demand time series in the national energy system model that will determine the required level of accuracy. Possible future work might be investigation of how the choice of heat demand method impacts the electricity demand time series generated from them and the resultant impacts on models of projected future peak loads and energy storage.

References

1. H. Lund, *Renewable Energy Systems: the Choice And Modelling of 100% Renewable Solutions* (Academic, Amsterdam, 2009)
2. H. Lund, F. Arler, F. Hvelplund, D. Connolly, P. Karnøe, Simulation versus optimisation: theoretical positions in energy system modelling. Energies, **10**(7) (2017). https://doi.org/10.3390/en10070840
3. S.D. Watson, K.J. Lomas, R.A. Buswell, Decarbonising domestic heating: What is the peak GB demand? Energy Policy **126**, 533–544 (2019). https://doi.org/10.1016/j.enpol.2018.11.001
4. R. Sansom, G. Strbac, The Impact of Future Heat Demand Pathways on the Economics of Low Carbon Heating Systems2. Presented at the BIEE – 9th Academic Conference, Oxford (2012)
5. L. Pedersen, J. Stang, R. Ulseth, Load prediction method for heat and electricity demand in buildings for the purpose of planning for mixed energy distribution systems. Energy Build. **40**(7), 1124–1134 (2008). https://doi.org/10.1016/j.enbuild.2007.10.014
6. A.J. Heller, Heat-load modelling for large systems. Appl. Energy **72**(1), 371–387 (2002). https://doi.org/10.1016/S0306-2619(02)00020-X
7. J. Love et al., The addition of heat pump electricity load profiles to GB electricity demand: Evidence from a heat pump field trial. Applied Energy **204**(C), 332–342 (2017). https://doi.org/10.1016/j.apenergy.2017.07.026
8. "ERA-Interim weather reanalysis. ECMWF. https://www.ecmwf.int/en/forecasts/datasets/reanalysis-datasets/era-interim
9. Eurostat population grid 2011. https://ec.europa.eu/eurostat/web/gisco/geodata/reference-data/population-distribution-demography/geostat. Accessed 2020
10. Energy Consumption in the UK end use tables
11. Gas Networks Ireland - Systems Performance Report 2018 (2018). https://www.gasnetworks.ie/corporate/gas-regulation/regulatory-publications/GNI-Systems-Performance-Report-2018.pdf
12. National Grid Gas Data Explorer. http://mip-prod-web.azurewebsites.net/DataItemExplorer

13. O. Ruhnau, L. Hirth, A. Praktiknjo, Time series of heat demand and heat pump efficiency for energy system modeling. Sc. Data **6**(1), 189 (2019). https://doi.org/10.1038/s41597-019-0199-y
14. A. Hooker-Stroud, P. James, T. Kellner, P. Allen, Toward understanding the challenges and opportunities in managing hourly variability in a 100% renewable energy system for the UK. Carbon Manag. **5**(4), 373–384 (2014). https://doi.org/10.1080/17583004.2015.1024955
15. J. Barton, R. Gammon, The production of hydrogen fuel from renewable sources and its role in grid operations. J. Power Sour. **195**(24), 8222–8235 (2010). https://doi.org/10.1016/j.jpowsour.2009.12.100
16. I. Staffell, S. Pfenninger, The increasing impact of weather on electricity supply and demand. Energy **145**, 65–78 (2018). https://doi.org/10.1016/j.energy.2017.12.051
17. Gas Demand Forecasting Methodology (2016). https://www.nationalgrid.com/sites/default/files/documents/8589937808-Gas%20Demand%20Forecasting%20Methodology.pdf
18. Python-statsmodels. https://www.statsmodels.org

Chapter 8
Detection of Patterns in Pressure Signal of Compressed Air System Using Wavelet Transform

Mohamad Thabet, David Sanders, and Nils Bausch

Abstract This paper investigates detecting patterns in the pressure signal of a compressed air system (CAS) with a load/unload control using a wavelet transform. The pressure signal of a CAS carries useful information about operational events. These events form patterns that can be used as 'signatures' for event detection. Such patterns are not always apparent in the time domain and hence the signal was transformed to the time-frequency domain. Three different CAS operating modes were considered: idle, tool activation and faulty. The wavelet transforms of the CAS pressure signal reveal unique features to identify events within each mode. Future work will investigate creating machine learning tools for that utilize these features for fault detection in CAS.

Keywords Compressed · Air · Systems · Intelligent · Wavelet

8.1 Introduction and Literature Review

This paper investigates detection of patterns in CAS pressure signal using a continuous wavelet transform (WT). CAS pressure signal characteristics that make WT a suitable analysis tool are discussed. Then, experiments performed and results obtained after applying WT are presented. Matlab function 'cwt' was used to apply WT. Results confirm that applying WT on a CAS pressure signal reveals unique patterns that can be used for fault detection.

Running a Compressed Air System (CAS) has a high energy cost [1, 2] and the efficiency of many CAS could be improved. Innovations in intelligent systems for automatic energy consumption control and fault detection might help address CAS energy efficiency [3].

Section 8.2 considers the CAS pressure signal, Sect. 8.3 presents the experiment and results and Sect. 8.4 concludes the paper.

M. Thabet (✉) · D. Sanders · N. Bausch
University of Portsmouth, Portsmouth PO1 2UP, UK
e-mail: mohamad.thabet@port.ac.uk

© The Author(s) 2021   61
I. Mporas et al. (eds.), *Energy and Sustainable Futures*, Springer Proceedings in Energy,
https://doi.org/10.1007/978-3-030-63916-7_8

8.2   Compressed Air Pressure Signal

A CAS pressure signal contains patterns that may be associated with operational events. Events, such as a compressor turning on or off in system with load/unload control, could be detected directly from the time domain pressure signal. However, detecting other events, such as tool or filter activation, was not as straightforward. In such cases, transforming the pressure signal to the frequency domain revealed features to recognise these events. The CAS pressure signal is non-stationary. For non-stationary signals, a Fourier transform provides information about frequency content, but not about time localization of those frequencies. A time-frequency signal processing tool, such as the WT, is more suitable for analysing a load/unload CAS pressure signal. A similar attempt to analyse a CAS pressure signal with WT was investigated in [4].

8.3   Experiments and Results

The industrial CAS installed in the University of Portsmouth was used for data collection. A pressure sensor was connected to the piping network and an air gun was used to simulate a tool activation. Data was recorded at a rate of 1 sample per second. The compressor had load/unload control, so the compressor turned on when the tank reached the lowest pressure limit and off when it reached the upper pressure limit. Three different scenarios were considered. The first corresponded to the case where no tool was activated (referred to as idle case). In the second case, an air gun was activated to simulate a tool activation. Finally, the third case corresponded to data recorded when the system had a leaking filter and different compressor control pressure limits. Each of these cases are analysed.

8.3.1   Idle Case

In the idle case compressed air consumption was due to small leaks in the system. Figure 8.1 shows pressure variation in this case. Pressure variation was obtained by removing the mean from the recorded data. The compressor switched on when the system pressure decreased by ~0.15 bar leading to a rapid increase in system pressure. Once the compressor switched off, the pressure decreased slowly. While the compressor was off, events such as filter activation, led to an increased loss in system pressure, reflected as a steeper decrease in pressure.

The WT of the idle signal is shown as a 3D contour plot in Fig. 8.2. On the z-axis, coefficient magnitude instead of coefficient are plotted to make visualization easier. Low frequency components (~0.01 Hz) had the highest coefficient magnitude and were present at all times. These components corresponded to the overall saw

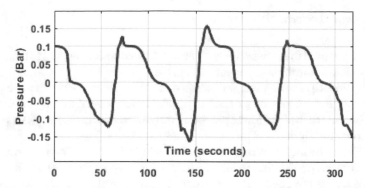

Fig. 8.1 Pressure signal change for the idle case

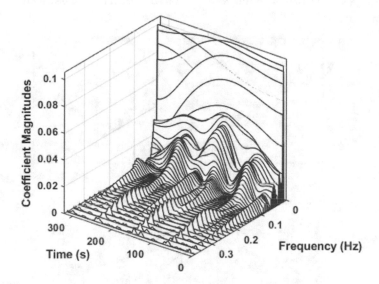

Fig. 8.2 Wavelet transform 3D contour plot for the idle case

tooth pattern of the signal (period of 100 s and frequency of 0.01 Hz). Compressor switching on introduced higher frequency components identifiable on the WT plot. These components had peaks (of different magnitudes) at frequencies of 0.05 and 0.25 Hz, both disappearing after the compressor switched off. Filter activation generated patterns with a peak at a frequency of 0.05 Hz.

8.3.2 Tool Activation Case

An air gun was used to study the impact of a tool activation. Compared to the idle case tool activation, while the compressor was on, led to longer compressor running times before the upper pressure limit was reached. Tool activation while the compressor was off increased the rate of air discharge, which accelerated pressure decrease so that the system reached the lower pressure limit faster. Figures 8.3 shows a pressure change while the air gun was on.

The WT of the tool activation signal is shown in Fig. 8.4. The plot reveals two patterns of interest. The first is the coefficients at frequencies of 0.05 Hz, and the second pattern is the coefficients at frequency range of 0.1–0.2 Hz. Both of these patterns were present at almost all times, unlike the idle case in which these frequencies only appeared when the compressor was on or when a filter was activated. While

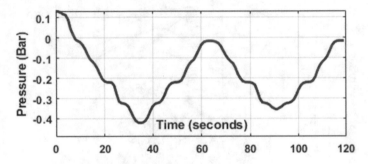

Fig. 8.3 Pressure signal change for tool activation

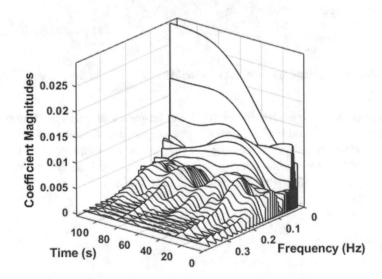

Fig. 8.4 Wavelet transform 3D contour plot for the tool activation case

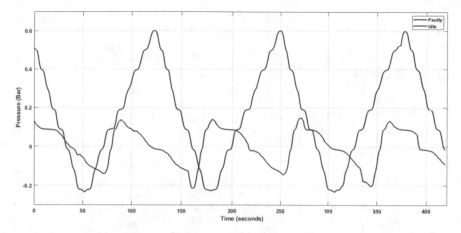

Fig. 8.5 Pressure change for a faulty system. The idle pressure signal shown in Fig. 8.3 is also shown for comparison

the tool was activated, compressor charging and discharging in the time domain signal had relatively similar slopes. This led to similar frequency components being present at all times.

8.3.3 Faulty Case

Data recorded when the system had two different faults was utilized to analyse faulty behaviour using a wavelet transform. The systems pressure control limits deviated from original settings, pushing the upper pressure limit higher. In addition, one of the filters had a relatively large leak. The pressure change for this case is shown in Fig. 8.5 alongside the idle case for comparison.

Because the upper pressure limit was higher, and due to the leak in the filter, the time the compressor spent on was longer. Moreover, the increased compressed air consumption due to the leak, meant the lower pressure limit was reached faster. The WT of the faulty system signal is shown in Fig. 8.6. Because the amplitude of the signal was relatively high (compared to idle and tool activation cases), the magnitude of the low frequency components was also high (0.14 compared to 0.025 and 0.1 in the idle and tool activation cases). The high frequency spectral components were similar to the ones seen for the tool activation case. This result is expected since a tool activation, such as an air gun, resembles the presence of a leak in the system.

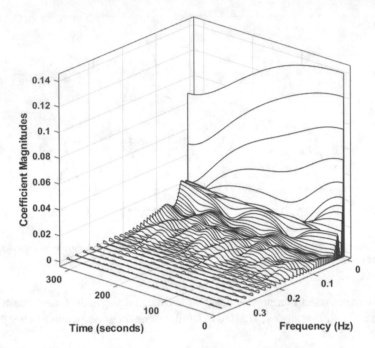

Fig. 8.6 Wavelet transform of a faulty system pressure signal shown in Fig. 8.5

8.4   Conclusion

This paper investigated the suitability of the WT for extracting features from the pressure signal of a CAS. Three different cases were considered: Idle, faulty and tool activation. The analysis proved that it is possible to generate features to identify different events, such as a tool activation and leaks. Future work will investigate employing those features to create a classifier for leak detection.

Acknowledgements The research was supported by the DTA3/COFUND Marie Skłodowska-Curie PhD Fellowship programme partly funded by the Horizon 2020 European Programme.

References

1. Fridén H, Bergfors L, Björk A, Mazharsolook E, Energy and LCC optimised design of compressed air systems: a mixed integer optimisation approach with general applicability. *Proceedings-2012 14th International Conference on Modelling and Simulatio* (2012), pp. 491–496
2. Murphy S, Kissock K, Simulating energy efficient control of multiple-compressor compressed air systems. *Proceedings of Industrial Energy Technology Conference* (2015)

3. M. Benedetti, V. Cesarotti, V. Introna, J. Serranti, Energy consumption control automation using artificial neural networks and adaptive algorithms: proposal of a new methodology and case study. Appl. Energy **165**, 60–71 (2016)
4. Desmet A, Delore M, Leak detection in compressed air systems using unsupervised anomaly detection techniques. *Proceedings of Annual Conference* of the *Prognostics and Health Management Society PHM* (2017), pp. 211–220

Chapter 9
The Impact of Data Segmentation in Predicting Monthly Building Energy Use with Support Vector Regression

William Mounter, Huda Dawood, and Nashwan Dawood

Abstract Advances in metering technologies and machine learning methods provide both opportunities and challenges for predicting building energy usage in the both the short and long term. However, there are minimal studies on comparing machine learning techniques in predicting building energy usage on their rolling horizon, compared with comparisons based upon a singular forecast range. With the majority of forecasts ranges being within the range of one week, due to the significant increases in error beyond short term building energy prediction. The aim of this paper is to investigate how the accuracy of building energy predictions can be improved for long term predictions, in part of a larger study into which machine learning techniques predict more accuracy within different forecast ranges. In this case study the 'Clarendon building' of Teesside University was selected for use in using it's BMS data (Building Management System) to predict the building's overall energy usage with Support Vector Regression. Examining how altering what data is used to train the models, impacts their overall accuracy. Such as by segmenting the model by building modes (Active and dormant), or by days of the week (Weekdays and weekends). Of which it was observed that modelling building weekday and weekend energy usage, lead to a reduction of 11% MAPE on average compared with unsegmented predictions.

Keywords Buildings · Data segmentation · Energy · Prediction · SVR

9.1 Introduction

With greater moves towards using machine learning in predicting building energy usage, there are many direct comparisons of learning techniques using the same data. However, it is considerably rarer when learning techniques are compared multiple times over multiple ranges [1–5]. With multiple ranges usually occurring when only testing a singular learning technique. Whilst a learning machine's accuracy can be

W. Mounter (✉) · H. Dawood · N. Dawood
Teesside University, Stephenson St, Tees Valley, Middlesbrough TS1 3BA, UK
e-mail: W.Mounter@tees.ac.uk

© The Author(s) 2021
I. Mporas et al. (eds.), *Energy and Sustainable Futures*, Springer Proceedings in Energy,
https://doi.org/10.1007/978-3-030-63916-7_9

observed when tested on a single dataset, the observed error is of limited value for comparative purposes with other learning techniques trained upon other data sets. As it is not known how each of the techniques would have performed if they had used the other test's data.

As such this study has been performed as part of a larger research project into investigating the impact of using different machine learning techniques (LR, SVR, ANN and Predictive Algorithms) on the accuracy of predicting building energy usage over a rolling horizon. In addition to investigating how the error that occurs in said predictions may be reduced through the use of data segmentation, such as if it is more accurate to model every single energy meter in a building and summate their predictions, or model the building as a whole. This specific paper focusing on the change in accuracy of building level energy use, by SVR (support vector regression), and the forecast range increase; and if error of long term predictions (Monthly) can be reduced to or bellow the level of the average error in short term predictions (Daily and weekly) through data segmentation.

Data segmentation, being the process of dividing and grouping data based on chosen parameters, in this case timeframes, so that it can be used more effectively; such as creating two separate prediction models for the separated groups. (As opposed to data splitting, in which data is randomly split for cross validation usage). Or to use an analogy, in cars, winter and summer tyres tend to perform better in their respective seasons than each other and all-season tyres, but poorer than each other and all-season tyres outside of their respective seasons. Or, in the case of machine learning, would a model trained with only with weekend building energy use data be more accurate at predicting weekend building energy use than a model trained with weekend and weekday energy use?

Additionally, Support Vector Machines (SVM) are supervised learning models with associated learning algorithms that evaluate data and identify patterns which can be used for regression analysis and classification; first identified by Vladimir Vapnik and his colleagues in 1992 [4, 7]. One of it's main advantages, and reasons for it's popularity in predicting building energy usage comes from it's ability to effectively capture and predict nonlinearity [6].

To investigate the study's aim the Clarendon Building, part of Teesside University Campus, was selected for use in this study- due to the data rich environment it's BMS (Building Management system) provided. Previous studies into this building utilizing square regression analysis typically had a baseline of "5% Mean Absolute Prediction Error (MAPE)" for the demands of each assets in one day ahead forecasts [8]. With investigations into the impact of data segmentation on ELM (Extreme learning machine) predictions using this data reduced the average of monthly prediction's percent error of the building's cooling system from 44.33 to 19.03%, reducing the error by 25.29% [9].

9.2   Research Method

From the Teesside university campus, amalgamated, building level and specific meter datasets were available of BMS data from January 2018 to December 2018. These datasets contained 15-minute averages of building elements energy usage, as well as sensory data of the internal and external environmental temperatures. From the Clarendon's BMS system, the total energy use of the building over the 2018 period, alongside their relevant timestamps, was extracted for use in investigating the accuracy of predicting building energy usage with support vector regression, over a variety of ranges and segmentations. Specifically:

- The impact of forecasting range on model accuracy, from one day to one month predictions into the future.
- The impact of segmenting the data along the lines of week days and week ends, creating independent prediction models for both, on the accuracy of said models compared with the base model.
- The impact of segmenting the data along the lines of active building periods and inactive periods, creating independent prediction models for both, on the accuracy of said models compared with the base model.

Noting that whilst the accuracy is expected to decrease as the forecast range increases, the point of interest of in studying the change of accuracy over a forecasting range, is the amount it increases by, and how significant that increase is.

To do this end, the data set was then segmented into it's individual seasons, before:

- The first month of each season, is placed into a season specific training data set.
- The second month of each season, is used to create three data sets that are to be used for prediction purposes, containing the first day, week and entirely of the month.

Creating in total four control training sets, and twelve control test data sets for evaluating each model's predictive accuracy over a rolling horizon. For the further segmented data, said control month (Training and testing) datasets would be subdivided into:

- Building active and inactive periods, where in separate models are created for the 8:00 am to 18:00 pm active period and the 18:15 pm to 7:45 am period of the data set.
- Weekdays and weekends, where in separate models are created for the Monday to Friday period and the Saturday to Sunday period of the data set.

Each model (A gaussian kernel SVR machine) is then trained with the first month of each season, to predict the following day, week and month of said season; to establish a baseline accuracy for each model. With the segmented data sets being trained and tested on their own relevant segments, the predicted values being recombined and then compared with the predictions made by the regression models trained on said months none-segmented data. Recording all of the predicted consumption for use in comparison with the actual consumption, evaluating the respective accuracies in terms of their Mean Absolute Percent Error.

9.3   Results and Discussion

9.3.1   *Daily, Weekly and Monthly Control Building Energy Predictions*

As can be seen in Fig. 9.1, as the forecast ranged increased so did the building energy use prediction error. But the increase in error was not proportional to the increase in forecasting range, e.g monthly predictions having thirty times the errors of daily ones. Instead, the bulk of the increase occurs between moving from daily predictions to weekly ones. Suggesting that the internal variations from the predicted 'norm' within a week are greater than the overall energy drift that occurs between the average energy use each week over the month period.

   An example of these variations from the predicted energy consumption can be observed when comparing the predicted energy consumption to the actual consumption, as can be seen in Fig. 9.2. The two dips in actual building energy use between event 200 and 400 of the first week of October 2018, can be observed in the majority of weeks through the year period, and are the result of reduced building activity during the weekend periods. As the prediction model has to accommodate for both the weekday and weekend energy use patterns, resulting in an average of the two, weighted in the weekday's favor, due to outnumbering the weekends.

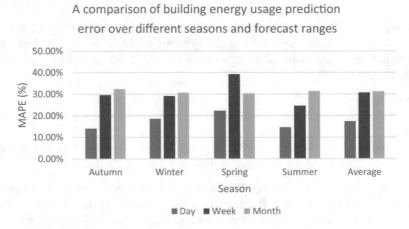

Fig. 9.1   A comparison of building energy usage prediction error

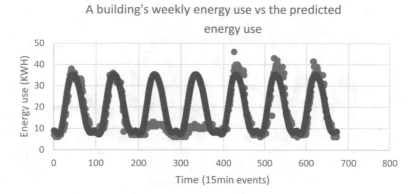

Fig. 9.2 The first week of autumn's actual vs predicted energy use

9.3.2   Segmented Monthly Building Energy Usage Predictions

As can be seen in Fig. 9.3, in every tested incent of weekday/end segmentation, the MAPE was reliable reduced comparatively to that of the unsegmented prediction, whilst the impact of segmenting by building active and dormancy periods had a negligible and erratic impact upon accuracy (<5%). This was unlike previous studies into the Clarendon building [9] where in segmenting between active and dormancy periods resulted in the greater reduction of prediction error. This is most likely due to the distinct difference in energy usage between the building's chillers being active and dormant, compared to the more continuous relationship and pattern of energy use observed at the summated level. (The Chillers being two separate patterns of energy use, whilst the total energy use being one (ignoring the weekday/weekend divide)).

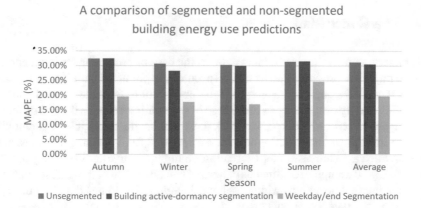

Fig. 9.3 A comparison of segmented and non-segmented building energy prediction error

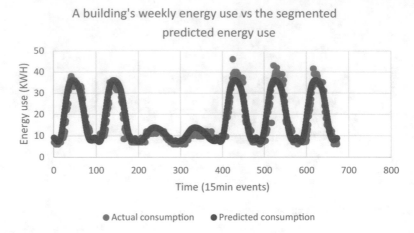

Fig. 9.4 The first segmented week of autumn's actual vs predicted energy use

As can be observed in Fig. 9.4, using multiple models to predict the week's building energy usage, allowed for greater accuracy in predicting the two distinct patterns of building energy use. Reducing the average MAPE of the monthly predictions from 31.18 to 19.73%, resulting in predictions with greater accuracy than the unsegmented weekly models (30.65%), having removed the main source of error that occurred in the weekly cycle. Though the segmented monthly prediction does possess 2.33% more MAPE than the average daily predictions (17.4%), that may potentially be attributed to the drift in energy usage from the training period from the changings of HVAC usage as weather patterns change with the progression of time, or from changes in overall building use from holidays, lesson scheduling and shifting building occupancy patterns.

9.4   The Conclusion

In summary, segmenting the training data, in the case of predicting building energy usage with SVRs, produces greater overall accuracy in predicted building energy use, when it is used to account for regular and reliable deviations from the 'normal' pattern of building energy usage. Such as in the case of segmenting between the two different building use patterns of weekends to week days. But having negligible to potentially negative impacts upon predictive accuracy when used to divide singular consistent patterns. Such as in the case of the Clarendon's buildings active and dormancy period, which followed a singular continuous pattern of behavior. Whilst the largest source of error (the internal deviations within each week period), could be addressed through segmenting the data, other data processing techniques or learning machines would have to be used to account for the drift in energy use that occurs over such long periods. As there is no distinct way to segment the training data to account for it.

Based upon these results, the three main areas of future work are as follows:

- The exploration of the data of the Clarendon building's specific meter energy consumption over the course of the year period, to observe how each building element's consumption changes over time.
- The comparison of the building energy model based upon the summated building energy use, with the summated predicted energy use of a set of models based upon each electric meter individually.
- The direct comparison of these results with the results of using the same datasets to train other learning machines previously mentioned.

Acknowledgements This research is a part of a PhD being undertaken at Teesside University under the supervision of Dr Huda Dawood and Prof. Nashwan Dawood. Additional acknowledgements should be given to Dr Chris Ogwumike for his assistance in acquiring the data used in this study.

Appendix

SVR calibration method, kernel selection and results, as well as other data created in this study, are available upon request.

References

1. M. Boegli, Y. Stauffer, SVR based PV models for MPC based energy flow management (2017)
2. Y. Chen, H. Tan, Short-term prediction of electric demand in building sector via hybrid support vector regression (2017)
3. Y. Chen, H. Tan, X. Song, Day-ahead Forecasting of Non-stationary Electric Power Demand in Commercial Buildings: Hybrid Support Vector Regression Based (2017)
4. Y. Ding, Q. Zhang, T. Yuan, Research on short-term and ultra-short-term cooling load prediction models for office buildings (2017)
5. S. Goudarzi et al., Predictive modelling of building energy consumption based on a hybrid nature-inspired optimization algorithm (2019)
6. M. Shen, H. Sun, Y. Lu, Household Electricity Consumption Prediction Under Multiple Behavioural Intervention Strategies Using Support Vector Regression (2017)
7. V. Vapnik, *The Nature of Statistical Learning Theory* (Springer, New York, 1995)
8. P. Boisson, S. Thebault, S. Rodriguez, S. Breukers, R. Charlesworth, S. Bull, I. Perevozchikov, M. Sisinni, F. Noris, M.-T. Tarco, A. Ceclan, T. Newholm, *DR Bob D5.1* (2017). https://www.dr-bob.eu/wpcontent/uploads/2018/10/DRBOB_D5.1_CSTB_Update_2018-10-19.pdf. Accessed 2019
9. W. Mounter, H. Dawood, N. Dawood, The impact of data segmentation on modelling building energy usage. In The International Conference on Energy and Sustainable Futures (ICESF). Nottingham Trent (2019)

Chapter 10
Development of an Advanced Solar Tracking Energy System

Samuel Davies, Sivagunalan Sivanathan, Ewen Constant,
and Kary Thanapalan

Abstract This paper describes the design of an advanced solar tracking system development that can be deployed for a range of applications. The work focused on the design and implementation of an advanced solar tracking system that follow the trajectory of the sun's path to maximise the power capacity generated by the solar panel. The design concept focussed on reliability, cost effectiveness, and scalability. System performance is of course a key issue and is at the heart of influencing the hardware, software and mechanical design. The result ensured a better system performance achieved. Stability issues were also addressed, in relation to optimisation and reliability. The paper details the physical tracker device developed as a prototype, as well as the proposed advanced control system for optimising the tracking.

Keywords Solar tracker · Physical design · Controller design · Stability · Optimisation

10.1 Introduction

Increased efforts in decarbonising the air includes the use of renewables, and coal's contribution to UK's energy capacity demands falling from 4% (2018) to just 2.8% (2019) is further confirmation of its downward trend. This is just one example of the effects that political decisions and commitments are having on our basket of energy sources contributing to the grid. The global uptake in renewables has incentivised research and development in energy capture, storage and application [1, 2]. The Solar Photovoltaic systems is one such technology that has grown vastly in its research and application [2]. UK alone has seen a growth of over 13 GW capacity in solar PV with deployment ranging from 4 KW to over 25 MW [3]. This paper focuses on optimising the Solar PV performance with respect to its power capacity output, and this is achieved through designing a tracker controller. The tracking system proposed here aims to verify and build upon other research efforts made in this field [4].

S. Davies · S. Sivanathan · E. Constant · K. Thanapalan (✉)
Faculty of Computing, Engineering and Science, University of South Wales, Pontypridd, UK
e-mail: Kary.thanapalan@southwales.ac.uk

© The Author(s) 2021   77
I. Mporas et al. (eds.), *Energy and Sustainable Futures*, Springer Proceedings in Energy,
https://doi.org/10.1007/978-3-030-63916-7_10

Whilst sub 4 KW systems make up over 90% of the number of installations in the first quarter of 2020 in the UK, it only amounts to 20% of overall capacity installed. The majority of the capacity is made up of installations greater than 5 MW [3]. Further research at the University of South Wales therefore hopes to apply the findings published in this paper to investigate the feasibility of deploying on a larger scale. This involves testing suitable methods of accomplishing a reliable way to monitor the production, and to balance the gains against the fact that we are now proposing to introduce moving parts into what is otherwise a mechanically advantageous static system. The tracking needs to increase the overall yield to be worthwhile, thus complementing the greater decarbonising effect of deploying on a larger scale.

10.2   System Configuration

Tracking the sun's path is one of the efficient measures that may be adopted to improve the panel performance. Several researchers have investigated many different tracking mechanisms [4, 5]. The physical solar tracking system construction (Fig. 10.1a, b) and its system performance depended on the choice of hardware, firmware and mechanical operation of the system. The system configuration described here is therefore with reference to its mechanical operation, and its hardware and firmware design. Initially a small-scale prototype system is investigated to serve as a proof-of-concept.

Mechanical operation of the system is designed to have the flexibility to rotate more than 270° azimuth, and ~90° in elevation. The stepper motor and the linear servo actuator are straight forward in their operation, but the linear actuator position should be designed carefully in order to achieve the maximum elevation of 90°.

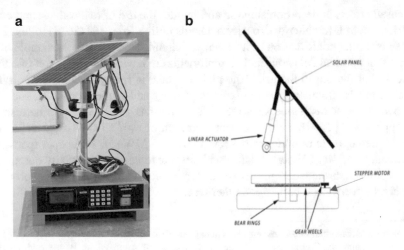

Fig. 10.1 a. Physical prototype-built. **b.** Mechanical model of the solar tracking system

The design was calculated taking into consideration the arrangement of suitable components which had been selected with reference to cost and performances. The design specification in this case was to achieve 90° in elevation. It was preferred to have a box arrangement under the mechanical system rather than leaving the base as plane metal. This improves the stability of the structure and secures some hardware components to sit within the box arrangement.

The hardware design of the system is mainly based on the M16C/62P microcontroller which improves system stability. Rotating mechanism is simplified in this work and the operation is simply interfaced to the Micro Controller Unit (MCU) using the most widely available and cost effective components in the market. This helps reduce the cost of the overall system prototype and better prepares this product for manufacturing. The solar tracker includes a microcontroller, and so firmware design is also an essential part of this system. Two possible design process was investigated in this work. They are firmware design for the microcontroller, and PC interface design [5]. The firmware design is a critical part of this system design and all the MCU configurations are reference to the datasheet of the M16C/62P. C language has been used to develop both Firmware for the MCU and Lab windows CVI 2010. Lab windows CVI has been used to develop the PC user interface development which based development environment offered by the National Instrument. The MCU firmware development structures are based on the state machine design adopted to superloop architecture. Each state is treated as separate functionality modules and can perform different functions until the state is changed or navigated through.

The remainder of this paper will describe the computer-based model developed to represent the small-scale solar tracking prototype already described (Fig. 10.1a, b). This model provides the basis for future math-based design, analysis and controller design of solar tracking energy system, to be applied for various applications including large scale deployment. The block diagram (Fig. 10.2) represent the solar tracker model, developed in MATLAB/Simulink™. The system has been developed from first principles for both the motor torque and the panel positioning. Panel weight and size, along with motor parameters, were selected to match as closely as possible

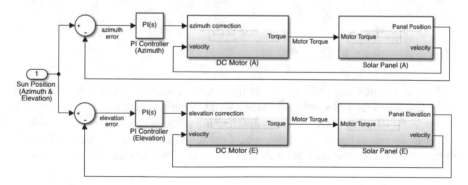

Fig. 10.2 Solar tracker system model

that of the physical prototype (Fig. 10.1a, b). The tracker system model receives a series of small step inputs representing the sun's azimuth and elevation path for the University's longitude and latitude, −3.33 and 51.59 respectively, for the complete year of 2020.

The system has poles that lie at the origin of the complex s-plane, but the feedback results in the torque output curve resembling an impulse response fed into the panel positioning, thus ensuring a stable system over all. The controller improves the system performance and serves to protect the motor's longevity over time by eliminating the otherwise overshoot during each step positioning update.

10.3 Controller Design

The target of the tracking control strategy is to develop a simple, but effective control method to obtain the desired positioning output with reference to sun position. This will ensure improved power capacity outputs. For simplicity and demonstration purpose a simple P-I (proportional integral) controller design was implemented (Fig. 10.2). A P-I controller is widely used in industrial control applications to regulate system variables. P-I controllers use a control loop feedback mechanism to control system variables and are both accurate and stable controllers. Without the controller at all, the system (Fig. 10.2.) is not responsive enough and has too large a settling time to cope with the physical dimensions modelled. The tuned controller on the other hand improves the settling time to under one second and does so without adding an overshoot or steady state error. The fourth-order system behaves in effect as a first-order system with the controller in place. Note, a differential term is not required due to there being no sudden changes on the input representing the sun's path, and so a two-term controller is sufficient in this case. The university is currently developing the system further so to be able to investigate the potential advantage of tracking stronger irradiance that may not always be on the sun's path. The controller in this case will need to be adapted to cope with more sudden changes with larger step movements.

10.4 Results and Discussion

Extensive research has concluded how tracking can improve the annual yield of a solar panel, and the proposed system here will look to further build on these results. The transfer function (1) for tracking the azimuth plane in Fig. 10.2 is shown below, where kd represent the damping constant, J the motor inertia, kf the back emf, kt the torque constant, L the motor inductance, and R the motor resistance.

$$\frac{ki}{s^4(R/L + kd/J)s^3(R * kd/L * J)s^2 + kp \cdot s + ki} \tag{10.1}$$

A comparison (Fig. 10.3a) of the system with and without the controller in place highlights the benefit of the controller. The step input represents a single elevation update of the panel. The controller dramatically improves the response rise time (s) and settling time (s), and does so without overshoot. This is even more critical if one wishes to update the panel position more frequently (e.g. each minute rather than hourly) as the panel elevation position may not have settled in time before receiving the next step input. The system is able to handle step inputs as frequent as 10 s intervals with the controller in place; evident in Fig. 10.3a. Of course, it is unnecessary to track the slow moving sun path so frequently.

A one-day snap-shot (Fig. 10.3b) serves as an example showing how the panel position successfully tracks the sun's path with accuracy. Here, the x-axis represent the time(s) from sunrise to sunset, and the y-axis represent the elevation in radians. The controller also tracks the sun's azimuth plane (Fig. 10.3c) as reliably as it does the sun's elevation, again over the same sun rise to sunset duration. The sun's azimuth is also recorded in radians, whereby 0 degrees radians in represents North. Based on the initial results collected to date, it is worth considering the impacts of scaling up this system. Several advantages and incentives will be further investigated, such as monetary gains from maximising governmental subsidies for on-grid applications,

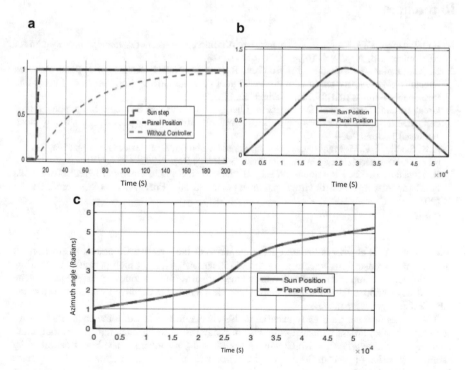

Fig. 10.3 a. Positioning with and without a controller. **b.** Controller tracking the Sun's elevation. **c** Controller tracking the Sun's azimuth plane

and technical gains in improving battery utilisation and battery sizing for on off-grid applications. There is also the environmental benefit in improving the carbon footprint, especially if scaled up to the larger solar farms.

10.5   Conclusion

The paper describes a possible solar tracking system that helps improve the power capacity generated. Suitable combination of hardware, software and mechanical design were detailed, and the results indicated the potential improvement in performance with a PI designed controller in place. The system design attempts to limit the costs whilst allowing for scalability. It's also worth pointing out that the benefits should also apply to off-grid systems, as tracking provides a higher/broader daily energy output, thus better utilising the battery's capacity, and so allowing for a greater battery capacity to be installed.

References

1. L.O. Polanco, V.M. Ramirez, K. Thanapalan, A comparison of energy management system for a DC microgrid. Appl. Sci. **10**(3), 1071–1074 (2020)
2. K. Thanapalan, E. Constant, M. Bowkett, Robust Controller Design for a Solar Powered Sustainable Energy System. In the Proceedings of the International Conference on Energy and Sustainable Futures (ICESF 19), Nottingham, UK (2019)
3. Great Britain. Department for Business, Energy and Industrial Strategy, Solar photovoltaic deployment (2020). https://www.gov.uk/government/statistics/solar-photovoltaics-deployment. Accessed 12 June 2020
4. A.K. Shukla, S.R.M. Manohar, C. Dondariya, K.N. Shukla, D. Porwal, G. Richhariya, Review on sun tracking technology in solar PV system. Energy Rep. 6:392–405 (2020)
5. S. Motahhir, A.E.L. Hammoumi, A.E.L. Ghzizal, A. Derouich, Open hardware/software test bench for solar tracker with virtual instrumentation. Sustain. Energy Technol. Assess. **31**, 9–16 (2019)

Chapter 11
Integration of Building Information Modelling and Augmented Reality for Building Energy Systems Visualisation

Vishak Dudhee and Vladimir Vukovic

Abstract Buildings consist of numerous energy systems, including heating, ventilation, and air conditioning (HVAC) systems and lighting systems. Typically, such systems are not fully visible in operational building environments, as some elements remain built into the walls, or hidden behind false ceilings. Fully visualising energy systems in buildings has the potential to improve understanding of the systems' performance and enhance maintenance processes. For such purposes, this paper describes the process of integrating Building Information Modelling (BIM) models with Augmented Reality (AR) and identifies the current limitations associated with the visualisation of building energy systems in AR using BIM. Testing of the concept included creating and superimposing a BIM model of a room in its actual physical environment and performing a walk-in analysis. The experimentation concluded that the concept could result in effective visualisation of energy systems with further development on the establishment of near real-time information.

Keywords Augmented reality · BIM · Buildings · Energy systems · Visualisation

11.1 Introduction

The buildings and construction sector accounted for 36% of final energy use and 39% of energy and process-related carbon dioxide (CO_2) emissions in 2018 [1]. In order to meet sustainability goals and mitigate climate change, energy systems must be more effective [2]. The visualisation of energy systems' performance can enhance the user's understanding of the system and, therefore, their decision-making relating to them [3]. Buildings consist of several energy systems, such as the heating, ventilation, and air conditioning (HVAC) systems, lighting systems and electric motors, out of which heating, space cooling, water heating, and lighting accounted for nearly 70% of site energy consumption [4]. The integration of Building Information Modelling

V. Dudhee (✉) · V. Vukovic
School of Computing, Engineering and Digital Technologies, Teesside University, Middlesbrough, UK
e-mail: V.Dudhee@tees.ac.uk

© The Author(s) 2021

I. Mporas et al. (eds.), *Energy and Sustainable Futures*, Springer Proceedings in Energy, https://doi.org/10.1007/978-3-030-63916-7_11

(BIM) and Augmented Reality (AR) provides the construction industry with many benefits, such as checking building designs [5, 6]. Although BIM models contain the energy systems information, the possibilities for visualising and analysing building energy system performance and behaviour by integrating BIM and AR have not been sufficiently explored.

11.2   Method

To explore the possibility of visualising digital data of energy systems and to understand the current limitations associated with the usage of BIM and AR for energy system visualisation, a BIM model was generated and visualised in AR using the process illustrated in Fig. 11.1.

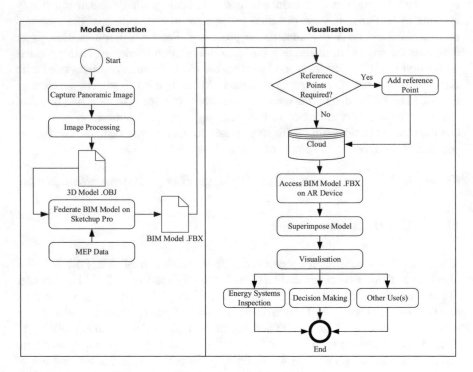

Fig. 11.1 BIM modelling and visualisation in AR schematic flowchart

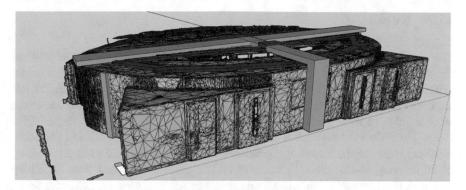

Fig. 11.2   Room mesh model with MEP components

11.2.1   Generation of a BIM Model

For the purposes of experimentation, a BIM model of a room at Teesside University with mechanical, electrical and plumbing (MEP) data was created, as shown in Fig. 11.2. An oval-shaped room was selected as atypical and particularly challenging for the traditional computer aided design software [7]. Therefore, the 3D model of the selected room was first automatically generated using a 3D camera (Matterport Pro 2) and Matterport app. SketchUp Pro was used to design and add MEP data and information to the model. The 3D model of the meeting room in OBJ format was imported and scaled 1:1 to avoid any issues that could arise related to scaling. A layer was then added for the room model, and MEP data were drawn and textured using three different layers and colours. The electrical lighting components are represented in red, the HVAC systems in grey and the piping system in green. Textual information was added near appropriate elements of the building to give further technical details of the systems.

11.2.2   Visualisation in AR

The generated model of the room was superimposed to the actual physical building using freely available software 3D Viewer Beta and visualised in Augmented Reality equipment Microsoft HoloLens 1–Developers Edition. The AR device's spatial mapping system generates a map of meshes that represent the physical space, which enables the placement of the digital model into the physical environment [8]. Using the 3D Viewer Beta AR software, the model was superimposed by trial and error. Since the software does not have a prescribed way of superimposing objects, the model was superimposed using arbitrary reference points. The appearance of physical systems found in the building was analysed to understand the possible uses of AR and current limitations in visualising energy systems.

11.3 Results

The room was visualised in AR using two different approaches: with and without the MEP systems added to the model. It was possible to successfully open the model using both approaches on AR software and interact with such models in the AR environment. The interaction and response times of the model were slow due to the large number of mesh elements it needed to process with the curved shape of the room. The different components of the room, such as the doors, windows and room furniture were visible; therefore, the different sections, such as the walls and ceiling, could be recognised. Although the model's texture was imported from SketchUp which showed colouring, the colour visualised in AR was plain white. There was a clear quality difference between the model in OBJ format and the model in the AR environment in FBX format. However, the model did show the actual geometry of the components, as presented in Fig. 11.3.

The MEP data were visible, and the system's components appeared to be on top of the ceiling. The electrical component added to the model logically matched the room lighting. The mechanical ventilation system was also positioned in a way so that it met the room ventilation cap and made its way out, passing through the ceiling and wall cavity. The water pipe, added as the plumbing system of the building, was also visible and located at the correct position. The written information about the different MEP systems and room components was visible and readable. Although the colour was not as assigned and presented in Fig. 11.2, the geometry was visible as modelled and positioned.

Fig. 11.3 Model visualisation in AR environment

11.4 Discussion

11.4.1 Visualisation Procedure

The visualisation procedure depends on various factors, such as the software used to create the BIM model, type of AR equipment and AR software being used. For this experiment, the BIM model created as described in Fig. 11.1 consisted of limited information, whereas having a more detailed BIM model could provide a better insight. The type of AR equipment used also had a significant impact on the AR experience. Mixed Reality (MR) headsets move with the user's head and do not limit the user from using their hands. In contrast, handheld AR equipment must be adequately positioned with the user's hands, potentially obstructing any simultaneous maintenance activities. The range of options available when interacting with BIM in an AR environment depends on the AR software being used. As a freely available software with limited features was used in the current study, it was not possible to view the texture or colour of the components in the model. Using this technique can be time-consuming and requires effort to achieve any level of accuracy in superimposing the model. This technique is not always feasible and can introduce alignment errors between the model and physical systems. Because the rotation and movement of the model in AR environment are completed by manipulating a sliding scale without defined values rather than with exact ratios, inaccuracies are common. For the model to be superimposed onto the physical building, the model needs to be aligned both horizontally and vertically using reference points and positioned on a horizontal flat surface before scaling can facilitate the superimposing process.

11.4.2 Current Limitations

A walk-through analysis of the model revealed that when the user moves through the space, the device loads only a particular section of the model within the user's field of view. Due to this loading process, inaccuracies in placement of model components were noticed during walk-throughs. For example, Fig. 11.4 shows the shifted position of the vertical pipes when viewed from different viewpoints in the room. Such imprecise positioning can result in the misinterpretation of building information.

Additional limitation is related to the type of data that could be viewed in AR, constrained to the geometrical design and pre-set information of the energy systems. The behaviour and performance of the system could not be identified or analysed in AR using the described approach. The presented approach is limited to visualisation of the static information of energy systems found in the BIM model. Not having the ability to visualise near real-time system information and the surrounding environmental information limits the possible uses of such visualisation systems. With

Fig. 11.4 Floating graphics

commercial usage of AR currently being under development, a limited variety of BIM-AR software is available, and none were identified as specialising in building energy system visualisation.

11.5 Conclusion

BIM models integrated with AR can be used to effectively visualise energy systems in a building, provided that the model contains the necessary and relevant information. The use of such a visualisation framework can allow for the analysis of building systems and the collection of necessary information found in the BIM model. BIM and AR's possible usage is still limited due to lack of real-time information and imprecise positioning during walk-throughs, which will be the subject of future work. The inclusion of near real-time data in BIM-AR visualisation techniques may improve understanding of system operation. Such integration framework will be explored in the future.

Acknowledgements The authors would like to acknowledge the help and financial support of Dr Kevin Thomas, Associate Dean International, School of Science, Engineering and Design, Teesside University who facilitated acquisition of Matterport services used in this study.

References

1. International Energy Agency, 2019 Global Status Report for Buildings and Construction. COP25. (UN Environment and the International Energy Agency, Spain, 2019)
2. H. Alemasoom, F. Samavati, J. Brosz, D. Layzell, EnergyViz: an interactive system for visualisation of energy systems. Vis. Comput. **32**, 403–413 (2016)
3. M.R. Herrmann, D.P. Brumby, T. Oreszczyn, X.M. Gilbert, Does data visualization affect users' understanding of electricity consumption? Build. Res. Inf. **46**, 238–250 (2018)
4. V.S. Harish, A. Kumar, A review on modeling and simulation of building energy systems. Renew. Sustain. Energy Rev. **56**, 1272–1292 (2016)

5. X. Wang, M. Truijens, L. Hou, Y. Wang, Y. Zhou, Integrating augmented reality with building information modeling: onsite construction process controlling for liquefied natural gas industry. Autom. Constr. **40**, 96–105 (2014)
6. C. Chai, K. Mustafa, S. Kuppusamy, A. Yusof, C.S. Lim, S.H. Wai, BIM integration in augmented reality model. Int. J. Technol. **10**, 1266–1275 (2019)
7. N. Kocic, Creating 3D model of architectural objects in different software. J. Ind. Des. Eng. Graph. **14**, 249–252 (2014)
8. G. Evans, J. Miller, M.I. Pena, A. MacAllister, E. Winer, Evaluating the Microsoft HoloLens through an augmented reality assembly application, in *Mechanical Engineering Conference Presentations, Papers, and Proceedings* 10197:0V1–0V16 (2017)

Chapter 12
GB Grid 9 August 2019 Power Outage and Grid Inertia

Christian Cooke

Abstract A power outage on 9 August 2019 raised questions about the ability of the GB electricity grid to withstand rapid changes in frequency caused by outages and surges on the network. Grid inertia has been changing in recent years due to the emergence of renewable energy generation as a significant contributor to the energy mix. Measures to mitigate this change need to be evaluated and the level of investment required to prevent a reoccurrence of such an event quantified. An outline is presented of a research programme towards this goal.

Keywords Power system dynamics · Inertial response · Frequency control

12.1 Introduction

12.1.1 An "Incredibly Rare Event" with "Immense Disruption"[1]

Lightning struck an overhead power line between Eaton Socon and Wymondley at 16:52.33 on Friday August 9, 2019.[2] Two major generation units went offline almost immediately (Fig. 12.1), followed by a cascade of outages that led to a cumulative power loss of 1,900 MW. This exceeded the capacity of the reserves held to maintain the integrity of supply. As a result, the power frequency, normally c. 50 Hz, dropped to 48.8 Hz (Fig. 12.2), triggering exceptional measures intended to preserve the stability of the overall network. 1.15 m households were disconnected, thousands of commuters were turned away from train stations, while hospitals and airports also suffered disruptions. It took 5 min for the frequency to recover to normal levels, and 45 min for all connections to be restored.

The event called into question the ability of the GB national grid to withstand rapid changes in frequency caused by outages and surges on the network. This *inertia* has been changing in recent years due to the emergence of renewable energy generation

C. Cooke (✉)
School of Mathematics and Statistics, The Open University, Milton Keynes, UK
e-mail: christian.cooke@open.ac.uk

I. Mporas et al. (eds.), *Energy and Sustainable Futures*, Springer Proceedings in Energy,
https://doi.org/10.1007/978-3-030-63916-7_12

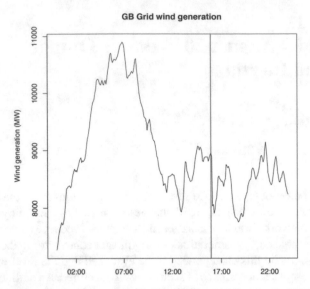

Fig. 12.1 Wind generation on August 9 2019, showing the drop in output of 737 MW from the Hornsea Windfarm (Data source: ElexonPortal.co.uk)

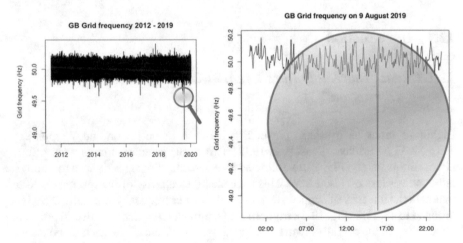

Fig. 12.2 Historical grid frequency, 2012–2019, August 9 2019 (Data source: https://gridwatch.templar.co.uk)

as a significant contributor to the energy mix. These power sources add to the power capacity of the network but not so much to inertia. A report into the event by Ofgem, [3] the energy regulator, recommended that the resilience of the grid should be examined in the context of this evolution of system inertia. Cost-effective measures to mitigate this change would need to be evaluated, quantifying the level of investment required to prevent a reoccurrence of an event such as the 9/8 outage.

12.1.2   Current Trends in Power Distribution

The increasing dependence on renewable energy to meet demand (Fig. 12.3) has been matched by the reduction in the number of fossil fuel-powered plants. Solid-fuel generation invariably involves the production of steam to drive a turbine, whose mechanical energy is converted into electrical energy. These turbines have significant inherent rotational inertia. Cumulative inertia across all turbines connected to the grid is synchronous and only small deviations from the overall rate are registered by individual turbines. Industrial motors directly connected to the grid similarly contribute to the system inertia (making up a significant proportion of demand-side inertia, which is responsible for up to 20% of the total in the case of the GB grid [4]). Solar and variable-speed wind power generation involves the generation of electricity directly and therefore make no contribution to the aggregate synchronous inertia.

A sudden drop in the aggregate power causes additional power to be drawn from connected power plants to make up the deficit, manifesting as an increased load, causing the turbine rotation to slow. The turbine inertia resists this change, with the rotational energy being converted to electrical energy compensating for the shortfall and the frequency of the AC power reduces. A similar situation occurs when there is a surplus of energy on the transmission network. As a result, power outages and surges in networks where there is a reduced level of system inertia carry the risk of rapid frequency changes. Frequency changes beyond permissible levels cause permanent damage and malfunction in AC motors connected to this supply.

It is clear that in the context of reduced systemic synchronous inertia, the consequences for grid management of the current and projected evolution of the power generation portfolio must be examined. Alternatives to alleviate the shortfall are the subject of active research: synchronous compensators, battery energy storage systems (BESS), synthetic inertia from wind turbines, and demand-side management. More accurate measurement of system inertia would also allow a more sophisticated response to network demands for inertia compensation.

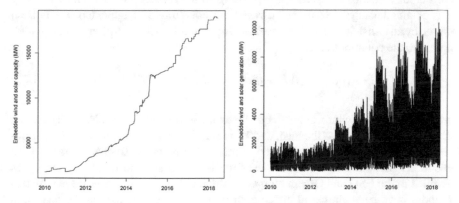

Fig. 12.3 GB grid renewable energy capacity and generation 2010–2018 (Data source: https://gri dwatch.templar.co.uk)

12.2   Investigations

12.2.1   Objectives

The focus of the project being undertaken is to design a model that permits an examination of the issues raised by the evolution of the energy mix on the national and international grid. To do this effectively the sophistication of the model must be balanced with its simplicity.

To fulfil these objectives the model should achieve a realistic overview of the impact of the evolution in generation mix, and consequent inertia profile, on grid stability. It should have the robustness necessary to identify situations leading to events such as that of August 9 and the flexibility to admit radical variations in energy mix under a variety of possible future scenarios.

An approach firmly rooted in nonlinear dynamical systems and control theory is intended, offering a continuous-time differential-equation model of the whole grid. To achieve this, an initial "toy" model is planned, using aggregated data grouped by the type of energy source, renewable (wind, solar), baseline (nuclear) and fossil fuel (gas, coal). The model is based on the stylized DC-Flow paradigm.

12.2.2   Initial Design to Study 9 August Outage

Demand is obtained from the 5-min Balancing Mechanism reporting, and this value is assigned to the single load bus. The frequency data used is the 1 s National Grid frequency database. Further refinement of the output would be obtained from a higher resolution data source which could be supplemented into the model without alteration.

Into this network we simulate the instantaneous frequency response by adjusting the output variables by levels corresponding to the recorded losses. The frequency f is obtained using the well-known swing bus equation [5], augmented with noise η driven by an Ornstein–Uhlenbeck process. In standard stochastic differential equation notation, the equations are.

$$df = (-\lambda_1 \Delta f - \lambda_2 \Delta P + \eta)dt, \qquad d\eta = -\eta/\tau dt + \sigma dW \qquad (12.1)$$

where Δf is the deviation of the frequency from 50 Hz, ΔP is the mismatch between the generated power and the load, and W is standard Brownian motion. The parameters λ_1 and λ_2 have been calibrated to match the recorded frequency trace of the event. ΔP is determined based on the timeline of losses and frequency responses detailed in the ESO and Ofgem reports on the incident [2, 3]. The noise parameters τ and σ have been chosen to match approximately the variations in the measured frequency trace.

12.3 Results and Discussion

The result of the simulation (Fig. 12.4) shows a correspondence between the frequency trace of the 9/8 event and the model. The model describes well the initial drop in frequency and its recovery, although it does not capture the subsequent overshoot in frequency after the losses are corrected.

We can use this correspondence to evaluate the effectiveness of the frequency response services deployed during the event, showing that the Control Room measures were more effective than LFDD (Low Frequency Demand Disconnection, where 1.15 m customers were put offline). Figure 12.5(i) shows that the simulated frequency trace returns close to 50 Hz without the LFDD intervention, whereas Fig. 12.5(ii) shows that without the Control Room actions the frequency falls further and remains below critical levels.

We can furthermore evaluate future frequency response technologies. The specification of the Dynamic Containment service to be launched by the ESO in Autumn 2020 [6] indicates that, to prevent a frequency drop of 1.2 Hz from an immediate

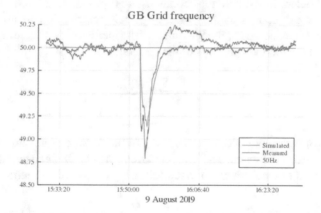

Fig. 12.4 Measured and simulated GB Grid

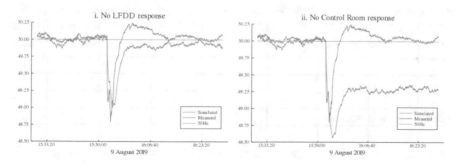

Fig. 12.5 August 9th simulation excluding LFDD (i) and Control Room (ii) responses

Fig. 12.6 Estimated
frequency nadir for ESO
Dynamic Containment
frequency response

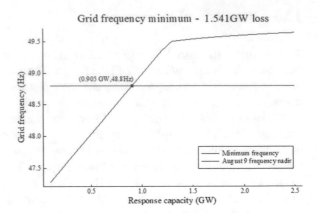

loss of 1.541GW, equivalent to that at the outset of the 9/8 event, a capacity greater
than 905 MW would be needed (Fig. 12.6).

The model can be improved by assigning sources of frequency response (battery,
pumped hydroelectric, synchronous compensators) to individual variables so as to
facilitate their calibration, thereby assessing their contribution to mitigating the
effects of significant losses of the 9/8 outage. The inertia of the system H_{sys} may be
obtained from the inertial constants of the generation sources.

$$H_{sys} = \frac{\sum_{i=1}^{N} S_i H_i}{S_{tot}} \tag{12.2}$$

$S_{tot} = \sum_{i=1}^{N} S_i$ is the sum of the rated powers of the N generators and H_i is the i th
generator's inertia coefficient. The simulated frequency is obtained by integrating an
aggregate model of the form given in the following coupled differential equations:

$$\frac{dG}{dt} = F_G(D, O, G, f) \qquad \frac{df}{dt} = F_f(D, O, G, f) \tag{12.3}$$

where G and f are the simulated aggregated conventional generated and system
frequency, D is the aggregate demand, O is system output comprising generation and
renewables, and the functions F_G, F_f couple G and f through the system mismatch
$D - O$.

12.4   Conclusion

Changing grid inertia is already a significant issue for grid operators and regulators.
The trend for reduced mechanical inertia from traditional sources will accelerate, as
renewables and interconnection form an increasing part of the energy mix. Measures

will need to be put in place to replace current sources of inertia in order to avoid the proliferation of further events comparable to the 9/8 incident in future. To further this goal, models will be needed that simulate rapid changes in frequency, the effects of grid inertia, and the effectiveness of proposed measures to mitigate adverse outcomes. These models will require a sophistication sufficient to provide a realistic evaluation of current and possible scenarios, and be adequately schematic so as not to require unnecessarily intense computational resources.

References

1. BBC News, UK power cut: National Grid promises to learn lessons from blackout (2019). 10/8/2019
2. ESO, Technical Report on the events of 9 August 2019 (2019). 6/9/2019
3. Ofgem, 9 August 2019 power outage report (2020). 3/1/2020
4. Y. Bian, H. Wyman-Pain, F. Li, R. Bhakar, S. Mishra, N.P. Padhy, Demand Side Contributions for System Inertia in the GB Power System. IEEE Trans. Power Syst. **33**(4), 3521–3530 (2018)
5. P. Vorobev, D.M. Greenwood, J.H. Bell, J.W. Bialek, P.C. Taylor, K. Turitsyn, Deadbands, Droop, and Inertia Impact on Power System Frequency Distribution. IEEE Trans. Power Syst. **34**(4), 3098–3108 (2019)
6. ES, Dynamic containment (2020). Retrieved from https://www.nationalgrideso.com/industry-information/balancing-services/frequency-response-services/dynamic-containment

Chapter 13
Analytical Model for Compressed Air System Analysis

Mohamad Thabet, David Sanders, and Victor Becerra

Abstract This paper presents a simple analytical model for a compressed air system (CAS) supply side. The supply side contains components responsible for production, treatment and storage of compressed air such as a compressor, cooler and a storage tank. Simulation of system performance with different storage tank size and system pressure set-point were performed. Results showed that a properly sized tank volume reduces energy consumption while maintaining good system pressure stability. Moreover, results also showed that reducing system pressure reduced energy consumption, however a more detailed model that considers end-user equipment is required to study effect of pressure set-point on energy consumption. Future work will focus on developing a supply-demand side coupled model and on utilizing model in developing new control strategies for improved energy performance.

Keywords Air · Modelling · Simulation · Energy

13.1 Introduction

Compressed air has been considered safe, easy to use, store and transport [1]. Because of these and other favourable characteristics, compressed air has been widely used in industrial plants [2]. CAS have been characterised by their low energy efficiency; 81% or more of energy supplied to CAS was wasted [3]. A modern CAS is formed of several sub-components [4], as shown in Fig. 13.1. It has been common to divide the compressed air system into a supply and a demand side. The supply includes components where compressed air was produced, treated and stored (compressor, cooler, filter, dryers, tank, etc.) while the demand side consisted of the distribution network and pneumatic devices.

Evaluating system performance through computer simulations can be effective for studying performance improvement measures. Friedenstein et al. presented in [5], a method to assess energy efficiency measures via computer simulation. The

M. Thabet (✉) · D. Sanders · V. Becerra
University of Portsmouth, Portsmouth PO1 2UP, UK
e-mail: Mohamad.thabet@port.ac.uk

© The Author(s) 2021   99
I. Mporas et al. (eds.), *Energy and Sustainable Futures*, Springer Proceedings in Energy,
https://doi.org/10.1007/978-3-030-63916-7_13

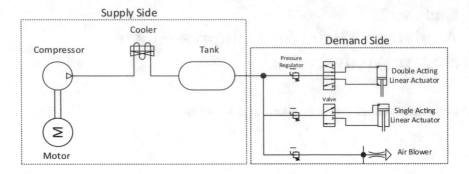

Fig. 13.1 System diagram

method had three main parts: system investigation, model creation and simulation. Maxwell and Rivera simulated in [6] the effect of different CAS control strategies on energy performance. Kleiser and Rauth studied in [7] system performance with different storage tank volumes was studied. In [8], Anglani et al. introduced a new tool that allowed a user to simulate different CAS components, configurations and settings.

This paper investigates a simplified model that can serve as a first tool for CAS design or retrofit evaluation. In Sect. 13.2, the mathematical models for the compressor, cooler and storage tank are presented. Section 13.3 evaluates, two different system modifications using the model: varying storage tank volume and decreasing system pressure. Finally, some conclusions and future work are presented.

13.2 Modelling Individual Components

13.2.1 Compressor

Assuming air behaved like an ideal gas, the work required (W_{comp}) to compress a volume (V_1) of air from air inlet pressure (P_1) to discharge pressure (P_2) was calculated using Eq. (13.1) [8].

$$W_{comp} = P_1 \times V_1 \times \frac{n}{n-1} \times \left[\left(\frac{P_2}{P_1} \right)^{\frac{n-1}{n}} - 1 \right] \quad (13.1)$$

where (n) is the polytropic compression exponent. The process was assumed to be isentropic and $n = 1.4$. To calculate the power, volume flow rate per unit time was used instead of volume. To estimate the power (W_{sup}) supplied to the compressor, efficiencies of the drive system (η_{ds}) and the compressor (η_c) were assumed constant at 90% and 80% respectively.

13.2.2 Air Cooler

In this study a counter flow air to air heat exchanger with effectiveness (ε) was assumed. The temperature of air leaving the cooler (T_3) was obtained using Eq. (13.2) [9].

$$T_3 = \varepsilon(T_{amb} - T_2) + T_2 \tag{13.2}$$

T_{amb} is the temperature of ambient air, which was assumed to be the cooling fluid. T_2 is temperature of air exiting the compressor and was obtained from ideal gas law.

13.2.3 Storage Tank

The purpose of a storage tank in a CAS was to store compressed air for when it was needed. Often, the pressure of air in storage was used as a control variable for the compressor. From the law of mass conservation, the mass of air in the storage tank was obtained with Eq. (13.3).

$$m(t) = \int_0^t (\dot{m}_{in} - \dot{m}_{out}) + m_0 \tag{13.3}$$

m_{in} and m_{out} were the air mass flowing in and out of the tank. m_0 is the mass of air in the tank at time $t = 0$. Assuming the temperature of air in the tank was equal to temperature of air leaving the cooler (T_3), the pressure of air in a tank (P_{tank}) of volume (V_{tank}) was obtained with Eq. (13.4) [6], where ($R = 287$ J/kg·K) is the gas constant of Air.

$$P_{tank} = \frac{m \times R \times T_3}{V_{tank}} \tag{13.4}$$

13.3 Simulation and Results

The mathematical model of the components presented in Sect. 13.2 were implemented in MATLAB. The compressor had a load/unload control, so the compressor would run at partial capacity (20%) for a period of time before shutting off. The system was assumed to leak 7% of its rated air capacity. System performance with no compressed air consumption, apart from leaks in the system, was simulated. Storage tank pressure is shown in Fig. 13.1. The load/unload control settings caused the tank pressure to cycled between the upper (9 bar) and lower (5 bar) pressure limits. In this case, consumption was only due to assumed leaks in the system.

13.3.1 Storage Tank Size

The impact of changing tank volume was studied. The compressed air consumption profile shown in Fig. 13.2 was assumed. Three different simulations with tank storage volumes of 100, 330 and 930 l were performed. The results compressor energy consumption and tank pressure are shown in Table 13.1 and Fig. 13.3 respectively. Results showed that a larger tank volume had a higher pressure stability, however this stability was not always justified in terms of energy consumption. The highest

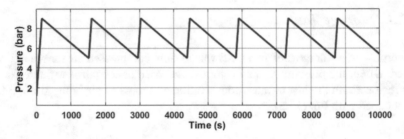

Fig. 13.2 Pressure of air in storage tank

Table 13.1 Energy consumption when different tank volumes were simulated	Tank volume (Litres)	100	330	930
	Energy Consumed (Kwh)	1.63	1.43	1.99

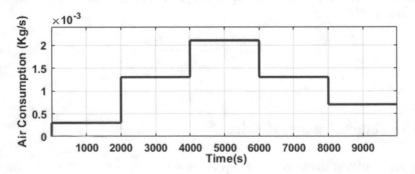

Fig. 13.3 Compressed air consumption

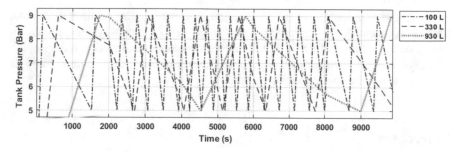

Fig. 13.4 Storage tank pressure when different tank volumes were simulated

Table 13.2 Energy consumption at different system pressure levels	System Pressure (Bar)	8	9
	Energy Consumed (Kwh)	1.33	1.43

energy consumption was for the 930 l tank, while the lowest was for the 330 l. Further analysis is required to study energy consumption for different compressed air consumption profile (Fig. 13.4).

13.3.2   System Pressure

Reducing system upper pressure limit from 9 to 8 bar was simulated assuming a tank volume of 330 l and the air consumption profile shown in Fig. 13.2. Energy consumption at different pressure levels is shown in Table 13.2. Results show that reducing system pressure led to 7% reduction in energy consumption. It should be noted that this model assumed constant compressor efficiency at different discharge pressures. In reality, efficiency varies with air discharge pressure. Moreover, leakage rate, which was assumed constant, changes proportionally with system pressure. Decreasing system pressure removes artificial demand in pneumatic tools. This could be better analysed through modelling the demand side of the system.

13.4   Conclusion

A simplified CAS model was presented. Mathematical expressions that describe compressor, cooler and storage tank were implemented in MATLAB. Different air storage tank volumes and different system pressure levels were simulated. Results for tank volume show that too large or too small a tank led to excessive energy consumption. An adequate tank volume reduced energy consumption while maintaining system pressure stability. Moreover, simulation results showed that reducing system pressure reduced energy consumption, however a more detailed model that considers demand side is required to properly analyse the effect of system pressure on energy consumption. Future work will investigate model validation and developing a supply-demand coupled model. Moreover, development of novel control strategies to reduce energy consumption will be studied.

References

1. T. Nehler, Linking energy efficiency measures in industrial compressed air systems with non-energy benefits–a review. Renew. Sustain. Energy Rev. **89**, 72–87 (2018). October 2017. https://doi.org/10.1016/j.rser.2018.02.018
2. R. Saidur, N.A. Rahim, M. Hasanuzzaman, A review on compressed-air energy use and energy savings. Renew. Sustain. Energy Rev. **14**(4), 1135–1153 (2010). https://doi.org/10.1016/j.rser.2009.11.013
3. M. Benedetti, F. Bonfa, I. Bertini, V. Introna, S. Ubertini (2018) Explorative study on Compressed Air Systems' energy efficiency in production and use: first steps towards the creation of a benchmarking system for large and energy-intensive industrial firms. Appl. Energy **227**, 436–448 (2018). June 2017. https://doi.org/10.1016/j.apenergy.2017.07.100
4. L. Berkeley, Compressed Air: a sourcebook for industry (2003), pp. 1–128
5. B. Friedenstein, J. van Rensburg, C. Cilliers, Simulating operational improvement on mine compressed air systems. South Afr. J. Ind. Eng. **29**(November), 69–81 (2018)
6. G. Maxwell, P. Rivera, Dynamic simulation of compressed air systems, in *2003 ACEEE Summer Study on Energy Efficiency* (2003), pp. 146–156
7. G. Kleiser, V. Rauth, Dynamic modelling of compressed air energy storage for small-scale industry applications. Int. J. Energy Eng. **3**(3), 127–137 (2013). https://doi.org/10.5923/j.ijee.20130303.02
8. N. Anglani, M. Bossi, G. Quartarone, Energy conversion systems: the case study of compressed air, an introduction to a new simulation toolbox, in *2012 IEEE International Energy Conference Exhibition ENERGYCON 2012*, (2012), pp. 32–38. https://doi.org/10.1109/EnergyCon.2012.6347776
9. T. Bergman, A. Lavine, F. Incropera, D. Dewitt, *Fundementals of Heat and MAss Transfer* (Wiley, 2011)

Chapter 14
Design Improvement of Water-Cooled Data Centres Using Computational Fluid Dynamics

Ramamoorthy Sethuramalingam and Abhishek Asthana

Abstract Data centres are complex energy demanding environments. The number of data centres and thereby their energy consumption around the world is growing at a rapid rate. Cooling the servers in the form of air conditioning forms a major part of the total energy consumption in data centres and thus there is an urgent need to develop alternative energy efficient cooling technologies. Liquid cooling systems are one such solution which are in their early developmental stage. In this article, the use of Computational Fluid Dynamics (CFD) to further improve the design of liquid-cooled systems is discussed by predicting temperature distribution and heat exchanger performance. A typical 40 kW rack cabinet with rear door fans and an intermediate air–liquid heat exchanger is used in the CFD simulations. Steady state Reynolds-Averaged Navier–Stokes modelling approach with the RNG K-epsilon turbulence model and the Radiator boundary conditions were used in the simulations. Results predict that heat exchanger effectiveness and uniform airflow across the cabinet are key factors to achieve efficient cooling and to avoid hot spots. The fundamental advantages and limitations of CFD modelling in liquid-cooled data centre racks were also discussed. In additional, emerging technologies for data centre cooling have also been discussed.

Keywords Data centre cooling · Computational fluid dynamics (CFD) ·
Turbulence modelling · Liquid–air heat exchanger

14.1 Introduction

The increase in the data centre industry in recent years along with High Power Computers (HPCs) energy consumption produced rapid growth on the server's heat density. Increasing numbers in data centre (DC) number and HPC's power densities is leading to an increase in high energy demand. The number of internet users is

R. Sethuramalingam (✉) · A. Asthana
Hallam Energy, Sheffield Hallam University, Sheffield, UK
e-mail: r.sethuramalingam@shu.ac.uk

© The Author(s) 2021
I. Mporas et al. (eds.), *Energy and Sustainable Futures*, Springer Proceedings in Energy,
https://doi.org/10.1007/978-3-030-63916-7_14

predicted to increase from 3.6 billion to 5 billion between 2018 to 2025 correspondingly, along with the number of internet of things (IoT) connections, expected to increase from 7.5 billion to 25 billion between 2018 to 2025 respectively [1]. In 2018, globally, data centre electricity consumption was estimated to be 198 TWh, which is almost 1% of global electricity demand [2]. Global data centre's energy consumption is predicted to increase by 15–20% per year in the future [3]. Cooling systems in data centres accounted for a major portion of the total data centre energy demand, which is estimated around 40% of its total energy consumption. Therefore, cooling systems effectiveness signifies the great window for the energy and cost saving. Most data centres are still using traditional air-cooled methods. However, these traditional air-cooling methods are inefficient and expensive to operate. Thus, alternative cooling methods should be considered to cope with rising cooling demand in data centres. Liquid cooling systems are one such solution which are in their early developmental stage. In this article, the use of Computational Fluid Dynamics (CFD) to further improve the design of liquid-cooled systems is discussed by predicting temperature distribution and heat exchanger performance.

HPCs are common in the recent data centres, where air-cooled systems might not be effective in cooling the electronic components of HPC. Therefore, liquid cooled systems are considered a promising solution for high heat density servers. Water-cooled systems have a higher heat transfer rate than the air-cooled systems due to the high specific heat capacity of liquid water [4]. This property of water allows it to operate effectively with lower temperature differences between server air and the coolant (water). Furthermore, water-cooled systems improve the effectiveness of heat removal and thus its energy efficiency by eliminating the need for a second heat exchange loop as required by traditional air-cooled systems [5]. Only a limited amount of published literature exists for liquid-cooled data centre systems, mainly focussing on direct liquid cooling with cold plate heat exchanger and micro channel flow [6–10]. This study primarily focuses on rear door liquid-cooled system with intermediate air-to-liquid heat exchangers, on which, there even less published literature.

In a liquid-cooled rear door heat exchanger rack, additional fans are placed in the rear door to overcome the greater pressure drop caused by the air-to-liquid cold heat exchanger. These fans draw in cold air from outside through the front door, transfer the heat from the servers to the air and then transfer the heat from the air to a liquid via the intermediate heat exchanger. In modern data centres, several individual server racks are packed into cabinets which have a combined heat density of 30 kW. Almoli et al., suggested that water cooled rear door coolers remove 90% of the server heat effectively with the remaining 10% lost to the surroundings through radiative transfer from the cabinet surfaces [11].

Other authors have attempted CFD modelling of heat transfer inside rack cabinets and could confirm their modelling results by validation with experimental results. A study by Almoli et al. accurately predicted air flow velocities and temperature distribution using ANSYS CFD software which was in close agreement with the experimental results [11]. In this current work, a similar study has been carried out using the ANSYS radiator model as air–water heat exchanger. The current work evaluates the cooling performance of the water-cooled rear door data centre rack at high heat density (40 kW) using similar validated CFD modelling procedure as used by Almoli et al. [11]. However, their results did not consider a radiator model to predict the actual performance of the heat exchanger. So, in this current work, simulations were carried out to evaluate the heat exchanger performance by using analytical calculations which were not considered in previous studies. In this current work, a similar numerical solver was be used with radiator model as air–water heat exchanger. The current work evaluates the cooling performance of the water-cooled rear door data centre rack at high heat density (40 kW) using similar validated CFD solver procedure but with different radiator model referred by Almoli [11].

14.2   Rack Cabinet Design

A typical 40 kW rack cabinet with rear door fans and an intermediate air–liquid heat exchanger is used in the CFD simulations to predict temperature distribution inside the cabinet and the heat exchanger performance. A 3-D model of the rack with rear door cooler has been modelled using ANSYS design modeller. The rack and heat exchanger have been constructed to represent 19-inch 42U data centre rack which has 40 kW cooling capacity. The height (Y), width (X) and length (Z) of the cabinet are 2 m, 0.6 m and 2 m respectively. The server rack is a framework where the server can be mounted. Racks usually have slots to place the servers in their designated place with the help of screws or bolts and nuts. 'U' refers to the standard to measure the vertical space in data server racks where 1 U = 1.75 inches. Figure 14.1 illustrates 3-D model of designed rear door unit with heat exchanger holder along with standard 42U data centre server rack and Fig. 14.2 shows the airflow direction as well as the server's position in the cabinet.

Fig. 14.1 Typical data centre cabinet and rear door cooler

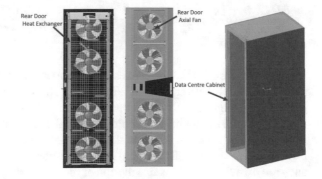

Fig. 14.2 Air flow
schematic through the water
cooled cabinet

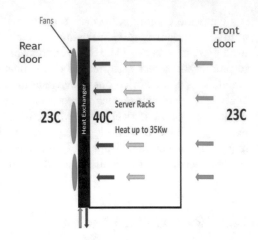

14.3 CFD and CAD Modelling

The air flow distribution and temperature difference within the cabinet were governed
by the conservation of mass, momentum and energy balance equations in the ANSYS
CFD solver. The radiator is placed inside the cabinet and it acts to extract the energy
from the servers' hot air flow. The CFD solver breaks down the original naiver stokes
equations into Reynolds averaged naiver stokes equations in order to solve the realistic engineering scenarios in reasonably less computational time [12]. Comprehensive description of the RANS equations can be found in the paper by Hussain et al.
[13].

For example, commonly used energy equation of the 3-D RANS model expressed
in the equation below, where "q" represents the heat source term and Re and Pr
represent Reynolds and Nusselt numbers respectively [14].

$$\frac{\partial T}{\partial t} + \frac{\partial}{\partial x}(uT) + \frac{\partial}{\partial x}(vT) + \frac{\partial}{\partial x}(\omega T) = \frac{1}{Re}\frac{1}{Pr}\left(\frac{\partial^2 T}{\partial x^2} + \frac{\partial^2 T}{\partial y^2} + \frac{\partial^2 T}{\partial z^2}\right) + q$$

(14.1)

The complete CFD Navier–Stokes solver consists of continuity, energy, turbulence
and momentum RANS equations. Widely used RANS turbulence model can be categorised into K-epsilon-RNG, K-Omega and Reynolds Stress model. The K-epsilon
model was used to minimise the computational time by the solver. Buoyancy effect
is used in CFD solver to simulate the air density change within the cabinet due to
heating. In this study, the Boussinesq approximation was used to account for the buoyancy force driving the convective motion of the fluid. The density is assumed constant
in the governing equations except in the buoyancy term (Boussinesq approximation).
The Computational Fluid Dynamics simulations were performed to investigate the
temperature distribution and cooling performance across the cabinet in the rear door

Table 14.1 Boundary conditions used in the ANSYS solver

Boundary condition	Symbol	Equation	Value
Inlet air flow velocity	U	–	m/s
Heat load	$\dot{Q}$	$\dot{Q} = \dot{m}Cp\Delta T$	40 kW
Inlet temperature	T-in	–	40°C
Outlet	P-out	–	Pressure outlet
Server temperature rise	ΔT	$\dot{Q} = \dot{m}Cp\Delta T$	16°C
Heat transfer coefficient	h	$\dfrac{\dot{m}c_p(T_{air,u}-T_{air,d})}{A} =$ $h(T_{air,d} - T_{ext})$	283.9 W/m^2.K
Heat exchanger loss coefficient	k_L	$\Delta p = k_L \frac{1}{2}\rho v^2$	18.6
Heat exchanger temperature	T-hx	–	14°C
Room temperature	T-room	–	25 °C

cooling configuration. Commonly used SIMPLE algorithm with higher order relaxation factor (0.75) is used in the numerical procedure to solve the K-epsilon RANS naiver stokes equation.

14.4   Boundary Conditions

The key properties to consider in a water-cooled data centre cabinet are (i) the temperature rise within the cabinet and (ii) heat exchanger properties. ANSYS fluent radiator model solves the transport equation which is a derivative of energy equation for heat removal. In addition, analytical heat transfer equations were used to estimate the heat exchanger performance to determine the cooling load of the cabinet. The temperature rise inside the cabinet is based on the heat load within the cabinet. This was determined by energy balance equation which given in Table 14.1. For this application, an air–liquid fin cooled heat exchanger was used with Darcy smooth approximation for pressure drop using Gnielinski correlation. The turbulence intensity was set at 7%. More comprehensive understanding of the pressure drop, air and liquid side heat transfer properties of the heat exchanger can be found in heat exchanger textbook [16]. Based on the initial analytical calculations, the following boundary conditions were used for the CFD simulations.

14.5   Results and Discussions

The post-processing results from CFD simulations are plotted below.

Figure 14.3 illustrates the iso-surface contours of the temperature upstream the heat exchanger, which indicates the temperature rise within the cabinet reaches the

Fig. 14.3 Iso-surface
temperature upstream the
heat exchange

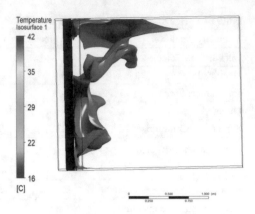

40 °C for 40 kW heat dissipating servers. Figure 14.4 illustrates the iso-surface
contours of the temperature downstream heat exchanger. The temperature contours
prove that the outlet temperature of the heat exchanger and room temperature are
same. This indicates that, 40 kW heat can be removed efficiently by using rear door
cooled heat exchanger without changing the room temperature, or in other words,
without the need for air conditioning.

Comparative results in numerical and analytical work reveal that air inlet temperature of the server cabinet plays a key role in the temperature rise inside the cabinet.
Similar performance was also spotted in the study by Almoli [11]. Figure 14.5 illustrates the temperature across the cabinet in mid Y axis plane upstream and downstream of the heat exchanger. It shows that the air is heated up to 40 °C at the inlet
and as it passes through the heat exchanger at the rear door, it cools down to the
target temperature which is about 23 °C, the room air temperature.

In Fig. 14.6, the red and blue lines represent the temperatures in the front and
back of the heat exchanger respectively. The results indicate that the hot air flow
temperature from the server (40 °C) has been cooled down to 24 °C after the heat
exchanger as it extracts the energy from the hot mainstream air. This indicates that

Fig. 14.4 Iso-surface
temperature downstream the
heat exchanger

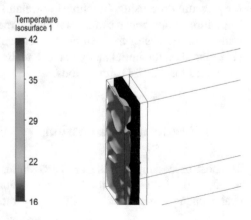

Fig. 14.5 Temperature
contours in mid plane

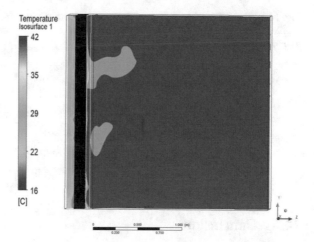

Fig. 14.6 Temperature
profile upstream and
downstream of the heat
exchanger

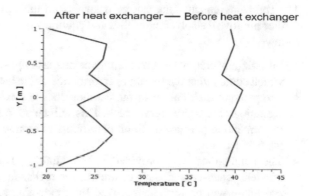

hot air from the server has been cooled down to the room temperature as it passes
through the heat exchanger. The quantitative data from two vertical lines further
illustrates the temperature profiles in Figs. 14.3, 14.4 and 14.5 Similar patters were
observed in a previous study [11].

In Fig. 14.7, the horizontal lines represent the temperatures in the front and the
back of the cabinet at three different heights. Results representing uniform temperature distribution can be obtained for a given air flow velocity and heat exchanger
parameters. Also, results indicate sudden temperature drop downstream the heat
exchanger in a uniform way, which ensures that temperature after the heat exchanger
reaches the room temperature for a given heat exchanger configuration. This eliminates the need for any external cooling or air conditioning in the data centre. The
results indicate that the temperature distribution predicted by the CFD solver can be
used as a benchmark for future energy efficient water-cooled data centre server rack
cabinets for further design improvements.

Fig. 14.7 Temperature
profiles through the cabinet
in horizontal planes

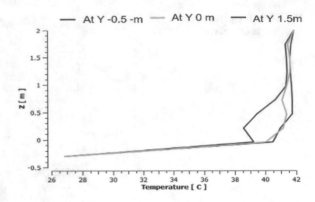

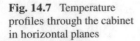

14.6 Conclusion

The following conclusions can be drawn from this study and the future work to
further improve the solver algorithm and accuracy of the simulation are identified as
follows.

- Inlet air temperature has great influence on the temperature rise within the cabinet,
  which could affect the thermal performance of the heat exchanger.
- Temperature distribution in the cabinet does not exactly represent the realistic
  scenario of the 40 kW server rack. This will be optimized in near future work by
  implementing porous condition paraments to represent the pressure drop across
  the cabinet.
- The simulations have provided a benchmark study of implementing energy
  efficient water-cooled data centre server racks. In the future, this simulation
  procedure will be further improved by conducting and validating experimental
  investigations.
- Finally, rather than imposing the Boussinesq approximation to simulation air
  density, the compressible flow will be considered in the near future numerical
  investigations.

Acknowledgements This work is supported by Innovate UK under KTP Programme No
KTP011150 and Impetus Enclosure Systems (Orion).

References

1. J. Koomey, Growth in data center electricity use 2005 to 2010 (2011). A report by Analytical
   Press, completed at the request of The New York Times.
2. E.R. Masanet, *Global Data Center Energy Use: Distribution, Composition, and Near-Term
   Outlook* (Evanston, IL, 2018).

3. K. Ebrahimi, G.F. Jones, A.S. Fleischer, A review of data center cooling technology, operating conditions and the corresponding low-grade waste heat recovery opportunities. Renew Sust Energy Rev **31**, 622–638 (2014)
4. Y.Q. Chi, J. Summers, P. Hopton, K. Deakin, A. Real, N. Kapur, H. Thompson, Case study of a data centre using enclosed, immersed, direct liquid-cooled servers, in *Proceedings of the 30th IEEE SEMI-THERM Symposium* (San Jose, CA, USA, 2014), pp. 164–173
5. M.M. Ohadi, S.V. Dessiatoun, K. Choo, M. Pecht, J.V. Lawler, A comparison analysis of air, liquid, and two-phase cooling of data centers, in *Proceedings of the 28th IEEE SEMI-THERM Symposium* (San Jose, CA, USA, 2012), pp. 58–63
6. S. Zimmermann, I. Meijer, M.K. Tiwari, S. Paredes, B. Michel, P.D. Aquasar, A hot water-cooled data center with direct energy reuse. Energy **43**(1), 237–245 (2012)
7. H. Coles, Direct Liquid Cooling for Electronic Equipment (2014)
8. https://www.asetek.com. (2020). Accessed March 2020
9. M. Iyengar, M. David, P. Parida, V. Kamath, B. Kochuparambil, D. Graybill, M. Schultz, M. Gaynes, R. Simons, R. Schmidt, T. Chainer, Extreme energy efficiency using water cooled servers inside a chiller-less data center, in *13th IEEE Intersociety Conference on Thermal and Thermomechanical Phenomena in Electronic Systems (ITherm)* (IEEE, 2012), pp. 137–149
10. Iceotope: https://www.iceotope.com (2020). Accessed 19 Mar 2020
11. A Ali (2013) Air flow management inside data centres University of Leeds, PhD thesis
12. B. Griffith, Q. Chen, Framework for coupling room air models to Heat Balance Model Load and Energy Calculations. HVAC&R Res. **10** (2) (2004)
13. S. Hussain, H. Patrick, Numerical study of buoyancy- driven natural ventilation in a simple three storey atrium building. Int. J. Sustain. Built Environ. **1**, 141–157 (2013)
14. H. Li, S. Lu, CFD Simulation of Computer Room Air Conditioning: Our 2D Navier-Stokes Solver coupled with Energy Equation vs. Fluent Simulation (2016)

Part III
Renewables

Chapter 15
Enhancing Methane Production from Spring-Harvested *Sargassum muticum*

Supattra Maneein, John J. Milledge, and Birthe V. Nielsen

Abstract *Sargassum muticum* is a brown seaweed which is invasive to Europe and currently treated as waste. The use of *S. muticum* for biofuel production by anaerobic digestion (AD) is limited by low methane (CH_4) yields. This study compares the biochemical methane potential (BMP) of *S. muticum* treated in three different approaches: aqueous methanol (70% MeOH) treated, washed, and untreated. Aqueous MeOH treatment of spring-harvested *S. muticum* was found to increase CH_4 production potential by almost 50% relative to the untreated biomass. The MeOH treatment possibly extracts AD inhibitors which could be high-value compounds for use in the pharmaceutical industry, showing potential for the development of a biorefinery approach; ultimately exploiting this invasive seaweed species.

Keywords Biofuel · Biogas · Seaweed · Post-harvest treatment

15.1 Introduction

Sargassum muticum is considered a menace as it is an invasive seaweed species to Europe [1]. Although this seaweed species has the potential to be exploited for its pharmacological and biomedical value, it is treated as 'waste' as there are no current utilisation methods. Methods to utilise this seaweed in a biorefinery approach could have positive economic and environmental implications.

Seaweed shows potential as a feedstock for biofuel production via methods such as anaerobic digestion (AD) that can handle wet biomass [1]. During AD, organic material is converted in an oxygen-free environment into CH_4 and carbon dioxide in a series of degradation steps carried out by different groups of bacteria and Archaea. However, the use of *S. muticum* as a feedstock for AD is currently limited by its low CH_4 yields [1]. This could partly be due to potential inhibitors of AD in seaweed; secondary metabolites extracted from *Asparagopsis taxiformis* showed inhibitory

S. Maneein (✉) · J. J. Milledge · B. V. Nielsen
Algae Biotechnology, University of Greenwich, Chatham Maritime ME4 4TB, UK
e-mail: sm8149w@gre.ac.uk

© The Author(s) 2021   117
I. Mporas et al. (eds.), *Energy and Sustainable Futures*, Springer Proceedings in Energy,
https://doi.org/10.1007/978-3-030-63916-7_15

effects against the ruminal microbial population and CH_4 production in the first 96 h [2]. The removal of other components of seaweed, such as polyphenols and salts, have also been associated with CH_4 yield enhancements [3].

This study aims to remove potential inhibitors of AD from *S. muticum* using aqueous methanol (MeOH). To the authors' knowledge, this is the first study to determine the biochemical methane potential (BMP) of aqueous MeOH treated *S. muticum* and subsequently, its prospective use as a feedstock for AD. The BMP of MeOH treated residues is also compared to washed seaweed, a pre-treatment method that enhances CH_4 production rates from summer-harvested *S. muticum* [4].

15.2 Experimental Methods

15.2.1 Sample Preparation

S. muticum samples were collected in May (spring) from Broadstairs, UK (TR399675). Samples of *S. muticum* were rinsed with deionised water (dH_2O) to remove sand and any residues from the seawater, stored at -18 °C, and freeze-dried (FD) (-55 °C, 48 h).

Three types of samples were prepared from spring-harvested *S. muticum* samples (Table 15.1). The MeOH treated residues (MTR) from three replicates were pooled and air-dried under the fume hood (24 h). For each replicate, the extracts from the sequential extraction were pooled and dried (Genevac™ Concentrator EZ-2). The wash solutions from the sequential washing were pooled and freeze-dried (-55 °C, 48 h), herein referred to as washed extract. Extraction yields were measured gravimetrically.

Table 15.1 Treatment methods of *Sargassum muticum* for BMP determination

Sample type	Treatment	Conditions
FD samples	Freeze-dried seaweed ground (Lloytron®, Kitchen Perfected coffee grinder) to a fine powder	–
Washed residues	5 sequential washes of ground FD spring samples (1:10 solid: solvent ratio) (n = 6)	FD samples mixed in deionised water, centrifuged (3,900 rpm, 20 min; Eppendorf, Centrifuge 5810R), repeated ×5
70% MeOH treated residues	3 sequential 70% aqueous MeOH (v/v) extractions of ground FD spring samples (1: 30 solid: solvent ratio) (n = 3)	FD samples mixed in 70% MeOH, incubated (room temperature (21.5 °C), 90 min), centrifuged (3,900 rpm, 20 min), repeated ×3

15.2.2   Dry Weight and Ash Content

Residues and extracts were dried in a vacuum oven at 105 °C overnight to determine their dry weight (DW) [4]. Ash and volatile solids (VS) contents were determined according to [5].

15.2.3   Biochemical Methane Potential (BMP) Determination and Specific CH_4 Yield Calculation

The inoculum was collected from an anaerobic digester treating paper-making waste at Smurfit Kappa Townsend Hook Paper Makers, Kent, United Kingdom. The inoculum was 'degassed' in a water bath (37 °C, 7 days) to minimise its contribution to the CH_4 yields during the BMP test [6], and then homogenised using a handheld blender (Philips™) before use.

The Automatic Methane Potential Test System II (AMPTS II) was used to measure CH_4 production. Replicates were made containing 2 g VS content of each sample type in Table 15.1. The inoculum was added to make an inoculum-to-substrate ratio of 5 and made up with water to 400 g. Blanks with only inoculum and water were used to calculate the net CH_4 production from *S. muticum*, removing the CH_4 contribution by the inoculum. Reactors were mixed continuously (150 rpm) and incubated at 37 °C. CH_4 volumes were recorded daily over 28 days and corrected to water vapour content at 0 °C, 101 325 kPa.

Specific CH_4 yields were calculated by multiplying the CH_4 potentials and the amount of residue remaining after the washing and extraction processes if 1 kg WW of *S. muticum* was treated according to Table 15.1.

15.2.4   Statistical Analysis

Excel (2016) was used for t-tests, and IBM SPSS version 25 was used for one-way and two-way ANOVA (with post-hoc analysis). Dependent variable: net cumulative CH_4 yield. Independent variables: treatment (FD samples, washed residues, MTR), day (time during BMP test).

Table 15.2 Volatile solids (VS) and ash content of residues and extracts as % of the dry weight (DW)

Samples	VS (% DW)	Ash (% DW)
FD *S. muticum*	74.7	25.3
Washed residues	85.5	14.5
Washed extract	60.1	39.9
MeOH treated residues	82.7	17.3
MeOH extract	64.9	35.1

15.3 Results

15.3.1 Dry Weight and Ash Content of Extracts and Residues

The material removed by washing FD *S. muticum* (42.8% DW) was significantly higher than those extracted by 70% aqueous MeOH (35.3% DW) (t-test, $p < 0.05$). The washed residues had a higher VS content and a lower ash content compared to MTR, corresponding to the higher ash content in the washed extract compared to the MeOH extract (Table 15.2). MTR and washed residues have significantly lower ash content and higher VS content relative to the FD samples (one-way ANOVA, $p < 0.05$).

15.3.2 CH₄ Production Profile and Specific CH₄ Yield

Figure 15.1 shows the CH_4 production profile of FD samples, washed residues and MTR during the BMP test. A significant interaction between days of incubation and

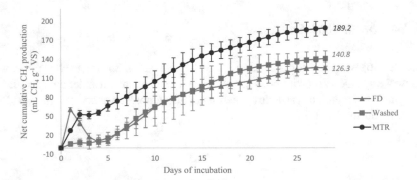

Fig. 15.1 Net cumulative CH_4 production of FD samples (FD [n = 3]; green triangles), washed residues (Washed [n = 6]; orange rectangles) and MeOH treated residues (MTR [n = 4]; red circles) over the duration of the BMP test (28 days). Final CH_4 potentials (after 28 days) are shown in italics in their respective colours. Error bars represent standard deviations

Table 15.3 Specific CH$_4$ yield of FD samples, washed residues and MeOH treated residues

Samples	Specific CH$_4$ yield (L CH$_4$ kg^{-1} WW *S. muticum*)
FD *S. muticum* (n = 3)	12.7 ± 0.9
Washed residues (n = 6)	9.1 ± 0.8
MeOH treated residues (n = 4)	14.3 ± 0.8

the treatment on the net cumulative CH$_4$ production was found (two-way ANOVA, $p < 0.05$). Post-hoc comparisons revealed that, except for day 2, the CH$_4$ production profile of MTR was significantly different from those of FD samples and washed residues for the duration of the BMP test ($p < 0.05$). In contrast, the CH$_4$ production profiles do not differ significantly between washed residues and FD samples ($p >$ 0.05), except for days 1 and 2 when the mean net cumulative CH$_4$ production by washed residues was significantly lower (up to 54.5 mL CH$_4$ g^{-1} VS) than FD samples ($p < 0.05$). Ultimately, while the final CH$_4$ potentials (after 28 days) of FD and washed residues were not significantly different ($p > 0.05$), those of the MTR's were statistically higher than both the FD samples and washed residues (one-way ANOVA, $p < 0.05$).

The specific CH$_4$ yield produced by MTR is statistically higher than the FD samples by 12.6% (t-test, $p < 0.05$) (Table 15.3). Comparatively, washing reduces the specific CH$_4$ yield by 28.3% relative to the FD samples; mean yields were statistically different (t-test; $p < 0.05$).

15.4 Discussion

Washing FD *S. muticum* removed 7.5% more compounds than 70% aqueous McOH. This is partly attributed to the higher ash content removed by water, which may be due to its higher solubility in water compared to MeOH. Higher extraction yields by water compared to alcoholic solvents are also found in the literature [7, 8]. The compounds removed by water had negative implications for AD of spring *S. muticum*. Despite enhanced VS content of 10.8% relative to the FD samples, washed residues still have lower specific CH$_4$ yields (3.6 L CH$_4$ kg^{-1} WW) and statistically similar CH$_4$ potentials to FD samples.

MTR showed higher VS content relative to FD samples, but lower VS content relative to washed residues. MTR produced 49.8% and 34.4% higher CH$_4$ potential relative to the FD and washed residues, respectively. The highest CH$_4$ production potential achieved by MTR could, therefore, be related to the removal of potential inhibitors of AD from *S. muticum*. MeOH extracts of *Sargassum spp.* showed antibacterial activity against *Bacillus subtilis*, with up to half the inhibitory potential of tetracycline, a potent antibiotic with anti-methanogenic activity [9, 10].

Other potential inhibitory compounds of AD include polyphenolic compounds and fatty acids [11]. However, polyphenolic and lipid content of the three sample types showed no correlation to the CH_4 potential (data not shown). Repeats of this experiment showed lower CH_4 potentials from MTR, but are nevertheless higher than FD samples. Higher CH_4 potentials in this experiment could be due to incomplete removal of aqueous MeOH. Further analysis of the antimicrobial potential and identification of compounds present in the aqueous MeOH extracts are needed to clarify their inhibitory potentials. The identification of compounds with a pharmacological or biomedical value may also allow for the development of a biorefinery approach.

15.5   Conclusion

Treating spring *S. muticum* with aqueous MeOH prior to AD can be an effective method to enhance CH_4 production, making it a more suitable feedstock compared to untreated *S. muticum*. This pre-treatment method shows potential in a biorefinery approach to ultimately exploit this invasive seaweed species when potential uses of compounds extracted by MeOH are identified.

Acknowledgements The authors acknowledge the financial support of the University of Greenwich and the DTA.

References

1. J.J. Milledge, P.J. Harvey, Ensilage and anaerobic digestion of Sargassum muticum. J. Appl. Phycol. **28**, 3021–3030 (2016)
2. B.M. Roque, C.G. Brooke, J. Ladau, T. Polley, L.J. Marsh, N. Najafi, P. Pandey, L. Singh, R. Kinley, J.K. Salwen, E. Eloe-Fadrosh, E. Kebreab, M. Hess, Effect of the macroalgae Asparagopsis taxiformis on methane production and rumen microbiome assemblage. Anim. Microbiome **1**, 3 (2019)
3. S. Maneein, J.J. Milledge, B.V. Nielsen, P.J. Harvey, A review of seaweed pre-treatment methods for enhanced biofuel production by anaerobic digestion or fermentation. Fermentation **4**, 100 (2018)
4. S. Maneein, J.J. Milledge, P.J. Harvey, B.V. Nielsen, Methane production from Sargassum muticum: effects of seasonality and of freshwater washes. Energy. Built. Environ. (In Press)
5. BSI, Solid biofuels-Determination of ash content. BS EN 147752009 (2009)
6. I. Angelidaki, W. Sanders, Assessment of the anaerobic biodegradability of macropollutants. Rev. Environ. Sci. Bio/Technol. **3**, 117–129 (2004)
7. S. Cho, S. Kang, J. Cho, A. Kim, S. Park, Y.-K. Hong, D.-H. Ahn, The antioxidant properties of brown seaweed (Sargassum siliquastrum) extracts. J. Med. Food **10**, 479–485 (2007)
8. I. Jerez-Martel, S. García-Poza, G. Rodríguez-Martel, M. Rico, C. Afonso-Olivares, J. Luis Gómez-Pinchetti, Phenolic profile and antioxidant activity of crude extracts from microalgae and cyanobacteria strains. J. Food Qual. **2017**(2017). Article ID: 2924508
9. I. Jaswir, A.H. Tawakalit Tope, R.A. Raus, H. Ademola Monsur, N. Ramli, Study on antibacterial potentials of some Malaysian brown seaweeds. Food Hydrocoll. **42**, 275–279 (2014)

10. J.L. Sanz, N. Rodríguez, R. Amils, The action of antibiotics on the anaerobic digestion process. Appl. Microbiol. Biotechnol. **46**, 587–592 (1996)
11. M. Czatzkowska, M. Harnisz, Inhibitors of the methane fermentation process with particular emphasis on the microbiological aspect: a review. Energy Sci. Eng. **8**, 1880–1897 (2020)

Chapter 16
Integration of Catalytic Biofuel Production and Anaerobic Digestion for Biogas Production

G. Hurst, M. Peeters, and S. Tedesco

Abstract The drive towards a low carbon economy will lead to an increase in new lignocellulosic biorefinery activities. Integration of biorefinery waste products into established bioenergy technologies could lead to synergies for increased bioenergy production. In this study, we show that solid residue from the acid hydrolysis production of levulinic acid, has hydrochar properties and can be utilised as an Anaerobic Digestion (AD) supplement. The addition of 6 g/L solid residue to the AD of ammonia inhibited chicken manure improved methane yields by +14.1%. The co-digestion of biorefinery waste solids and manures could be a promising solution for improving biogas production from animal manures, sustainable waste management method and possible form of carbon sequestration.

Keywords Biorefinery · Levulinic acid · Solid residue · Anaerobic digestion · Hydrochar

16.1 Introduction

Lignocellulosic biomass is the most abundant renewable feedstock for biofuel production globally [1]. Biorefineries utilise a range of thermochemical processes to produce high platform chemicals for fuels and niche applications. Acid hydrolysis has received significant interest recent years for the valorisation of cellulose and hemi-cellulose fractions of biomass. Acid catalysts such as sulphuric acid (H_2SO_4) and hydrochloric acid (HCl) under aqueous conditions at mild temperatures (160–250 °C) can produce levulinic acid, lactic acid and 5-hydroxymethfural (5-HMF) [2]. Of which levulinic acid has been recognised as one of the most promising precursors for catalytic biofuel production. Several commercial pilot plants by GFBiochemicals

G. Hurst (✉) · S. Tedesco
Department of Mechanical Engineering, Manchester Metropolitan University, Dalton Building, Chester Street, Manchester M1 5GD, UK
e-mail: george.hurst2@stu.mmu.ac.uk

M. Peeters
School of Chemical Engineering, Newcastle University, Newcastle upon Tyne N1 7RU, UK

© The Author(s) 2021
I. Mporas et al. (eds.), *Energy and Sustainable Futures*, Springer Proceedings in Energy,
https://doi.org/10.1007/978-3-030-63916-7_16

(Caserta, Italy) and Segetis (USA) have demonstrated the potential technologies for the large scale deployment of catalytically produced biofuels [3].

The growth of biorefineries will create challenges to valorise all waste products. Acid catalysis of lignocellulose produces significant amounts of hydrochar-like Solid Residue (SR) material. Recent studies have utilised SR as a solid fuel, pyrolysis feedstock and building material among other applications [4]. Hydrochar has recently been investigated as Anaerobic Digestion (AD) supplement [5]. AD is a sustainable low-cost disposal process for organic matter as well as a source renewable energy in the form of biogas. Biogas is considered an important low-carbon bioenergy source however further implementation is limited due to feedstock constraints. Animal manure is among the most pressing environmental concerns and potential AD feedstock but is limited due to high inhibitor concentrations, most notably ammonium [6]. Recently, hydrochar has been shown to improve biogas yields from swine manure by 32–52% [7] by adsorbing ammonium, promoting microbial growth and buffering capacity during AD; but is limited due to the high costs associated with hydrochar production.

The pseudo-hydrochar properties of SR from acid catalysis could potentially be used as a low-cost hydrochar for anaerobic digestion supplementation, in order to improve the economics of both levulinic acid production and biogas. In this study, we investigated the feasibility of using SR from the sulphuric acid hydrolysis of *Miscanthus x Giganteus* to improve the anaerobic digestion of chicken manure.

16.2   Experimental

16.2.1   Solid Residue Preparation and Characterisation

Miscanthus x Giganteus was dried, ball-milled (<0.2 mm) and stored in air tight conditions for further use. Solid residue was prepared by heating 0.5 g of biomass in 10 ml of 2 M H_2SO_4 catalyst at 180 °C for 60 min under microwave heating. Post reaction the SR was separated, washed with deioinsed water and dried at 60 °C. The post aqueous reaction media was analysed using an HPLC as previously reported [8]. Levulinic acid and SR yields are reported on a dry mass basis and theoretical levulinic acid yield was calculated on structural sugar basis determined according to NREL standard 510-42618. The acid hydrolysis process yielded; 30.8 wt.% SR and 16.7 wt.% levulinic acid which corresponded to a theoretical 64.5% levulinic acid yield.

The total solids content and ash content were determined according to standards ASTM D4442 and NREL 42622 respectively. The CHNSO elemental combustion was carried using an Elemental Vario Macro Cube analyser, with oxygen % calculated by difference. The recalcitrance index, R_{50}, for the SR was calculated according to Harvey et al. to estimate the carbon sequestration potential [9].

16.2.2 Batch Anaerobic Digestion

Chicken manure (CM) was collected from a local supplier, dried and stored in a desscitator until use. Inoculum sludge was collected from a local mesophilic AD plant operating at 31 °C. The total volatile solids of CM and inoculum was set at 8 wt.% with a C/N ratio of 7.1. The effect of SR on AD was investigated by varying the SR concentration by 2–10 g/L, compared with an un-supplemented control reactor. The biomethane potential assays were conducted in 500 ml glass flasks with working volume of 200 ml and purged with nitrogen gas for 5 min to achieve anaerobic conditions. The batch reactors were submerged in a water bath maintained at 31 ± 1 °C and the manually shaken once a day. The batch biomethane potential (BMP) assays were set up in duplicate and operated for 14 days with daily gas measurements. The primary biogas components, CH_4 and CO_2, were analysed utilising a GeoTech 2000 biogas analyser. The cumulative biogas volumes were then fitted to the modified Gompertz equation shown in Eq. 16.1 [6], using non-linear regression analysis in Matlab© (2016a).

$$V_{CH4}(t) = A_{max}\exp\left[-\exp\left(\frac{R_{max} * e}{A_{max}}(\lambda - t) + 1\right)\right] \qquad (16.1)$$

16.3 Results

16.3.1 Material Characterisation

The properties of SR, Inoculum and CM are shown in Table 16.1. The SR carbon content exceeded 62%, with a high H/C and low O/C ratio indicating that the biomass underwent dehydration during the acid hydrolysis process, to produce SR with similar elemental properties to hydrochar. The acid hydrolysis process resulted in a high specific surface area (19 m^2/g) which suggests the SR contains micro-porous structures suitable for adsorption. The pH of SR was measured in water at 1:10 mass ratio and the solid residue was slightly acidic and could negatively affect buffering capacity of AD. The recalcitrance index, $R_{50,}$ is a measurement of char stability to microbial degradation and suitability for carbon sequestration. The SR had a R_{50} of 0.66 which is comparable with that of biochar and indicates the land application of post-digestion SR could be a potential form of stable carbon sequestration.

Table 16.1 Characteristics of inoculum, chicken manure (CM) and SR

	Raw-Miscanthus	SR	Inoculum	CM
TS (%, w.b)	97.2	95.15	7.6	87.7
VS (%, w.b)	94.4	92.4	5.5	64.7
VS (%, d.b)	97.37	97.37	72.1	73.7
pH	3.56	2.5	7.78	N/A
BET (m^2/g)	N/A	19	N/A	N/A
C (%, DS basis)	48.1%	62.4%	32.1	34.7
H (%, DS basis)	6.0%	5.5%	4.7	5.2
N (%, DS basis)	0.5%	0.3%	7.2	7.7
S (%, DS basis)	0.1%	0.4%	0.9	0.8
O (%, DS basis)	42.4%	23.8%	27.1	23.8
O/C	N/A	0.29	N/A	N/A
H/C	N/A	1.06	N/A	N/A
R_{50}	N/A	0.66	N/A	N/A

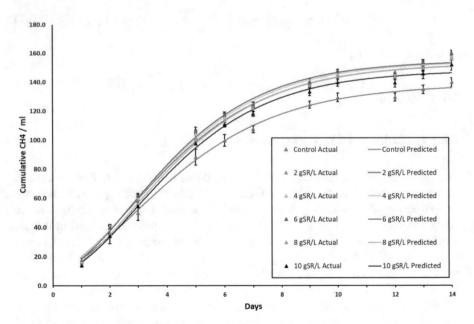

Fig. 16.1 Cumulative methane yield with varying SR addition and predicted yields

16.3.2 Effects of SR on Methane Production During CM Digestion

Figure 16.1, shows the cumulative methane yields from the AD of CM over 14 days, alongside the predicted methane yields according to Eq. 16.1. The control reactor yielded 136 ml CH_4/gVS compared with the CM stoichiometric CH_4 potential of 286 ml/gVS, which corresponds with a biodegradability index of 48%. The addition of 2 gSR/L increased the methane yield (+12% CH_4) with respect to the CM control, strongly indicating a beneficial effect to AD, by either promoting microbial growth or adsorbing inhibitors. There was only a minor improvement with increasing SR concentration at 6 g/L (+14.1% CH_4) suggesting that the SR is not being degraded in the reactors. The minor improvement also indicates that the AD system is not undergoing significant stress and the positive effects of SR could potentially be more significant in more toxin-rich environments. There was a noticeable decrease in CH_4 yields at higher SR loads (8 and 10 gSR/L), which can be attributed to the potential toxicity of SR on AD microorganisms. However, for all BMP conditions, SR addition resulted in higher methane yield than the CM control, which emphasises the positive effects on biogas production.

The Gompertz model has successfully been used to model BMP experiments from a range of substrates [6, 10]. For all experimental conditions in this study, the regression values exceeded 0.99, indicating a good fit of the model parameters shown in Table 16.2. The addition of SR increased the maximum methane production (A_{max}) by 6.9–11.8% and the maximum methane production rate (R_{max}) by 18.3–25.9%, compared with the control reactor. This suggests that SR increased both the biodegradability of CM over 14 days and also the maximum microbial activity. The lag time of microbial AD system is associated with the maximum degradation rate and the adaption of the microbial community to the reaction conditions. Table 16.2, shows that SR concentrations prolonged the lag phase compared with the control

Table 16.2 Summary of Kinetic data for the AD of CM with different SR concentrations

Condition	Cumulative CH_4 yield	Modified Gompertz parameter			Statistics	
	F	A_{Max}	R_{Max}	λ	R^2	p
Control	140 ± 3	138.4	18.44	0.1696	0.9960	0.9976
2 g/L	156 ± 2	152.7	21.7	0.256	0.9946	0.9966
4 g/L	158 ± 2	154.4	22.51	0.3189	0.9937	0.9942
6 g/L	160 ± 1	154.8	23.22	0.3974	0.9917	0.9939
8 g/L	157 ± 1	152.4	22.32	0.4615	0.9949	0.9960
10 g/L	152 ± 3	148.0	21.81	0.492	0.9945	0.9963

F is the measured cumulative methane production, mL/gVS added, A_{max} is the predicted cumulative methane production, mL/gVS added day, R_{Max} is the maximum methane production rate, mL/gVS day, and λ is the duration of the lag phase, days

reactor from 0.17 to 0.49 days. The increase in lag time implies a mild microbial inhibition with increasing SR concentrations and that the microbial community reacted negatively to those SR levels. This can be further seen with a decrease in A_{max} from 154.8 to 148.0 ml CH_4/gVS day between 6 and 10 g/L suggesting that inhibition is caused from the SR itself and not from imbalances in the complex system. The negative effects of SR on AD must be further investigated, but does not negate the potential benefits.

16.4 Conclusion

In this study, the addition of SR from acid hydrolysis improved the methane yields, by 6.9–14.1%. During the mesophilic anaerobic digestion of CM. The SR from a high yielding levulinic acid process was determined to have hydrochar properties which were evident during the AD process. The modified Gompertz model suggested that SR increased both the microbial degradation rate and the cumulative methane yields, however toxicity was a factor at higher concentrations. The integration of AD into acid hydrolysis biorefineries could potentially create synergies for the dual production of biofuel and biogas.

Acknowledgements This research was supported by Manchester Metropolitan University via the Research Accelerator Award (RAG 2017/18-113380). BBSRC NIBB's network High Value from Plants funded the training and compositional analysis of biomass in conjunction with Celignis ltd. (CORE-TA–19).

References

1. S. Takkellapati, T. Li, M.A. Gonzalez, An overview of biorefinery-derived platform chemicals from a cellulose and hemicellulose biorefinery. Clean Technol. Environ. Policy **20**(7), 1615–1630 (2018)
2. A. Mukherjee, M.J. Dumont, V. Raghavan, Review: sustainable production of hydroxymethyl-furfural and levulinic acid: challenges and opportunities. Biomass Bioenerg. **72**, 143–183 (2015)
3. F.D. Pileidis, M.M. Titirici, Levulinic acid biorefineries: new challenges for efficient utilization of biomass. ChemSusChem **9**(6), 562–582 (2016)
4. S. Kang, J. Fu, G. Zhang, From lignocellulosic biomass to levulinic acid: a review on acid-catalyzed hydrolysis. Renew. Sustain. Energy Rev. **94**, 340–362 (2018)
5. W. Guo, Y. Li, K. Zhao, Q. Xu, H. Jiang, H. Zhou (2019) Performance and microbial community analysis of anaerobic digestion of vinegar residue with adding of acetylene black or hydrochar. Waste and Biomass Valoriz., no. 0123456789
6. G.K. Kafle, L. Chen, Comparison on batch anaerobic digestion of five different livestock manures and prediction of biochemical methane potential (BMP) using different statistical models. Waste Manag. **48**, 492–502 (2016)
7. H. Jin et al., Hydrochar derived from anaerobic solid digestates of swine manure and rice straw: a potential recyclable material. BioResources **13**(1), 1019–1034 (2018)

8. G. Hurst, I. Brangheli, M. Peeters, S. Tedesco, Solid residue and by-product yields from acid-catalysed conversion of poplar wood to levulinic acid. Chem. Pap. **74**(5), 1647–1661 (2020)
9. O.R. Harvey, L.J. Kuo, A.R. Zimmerman, P. Louchouarn, J.E. Amonette, B.E. Herbert, An index-based approach to assessing recalcitrance and soil carbon sequestration potential of engineered black carbons (biochars). Environ. Sci. Technol. **46**(3), 1415–1421 (2012)
10. J. Pan, J. Ma, X. Liu, L. Zhai, X. Ouyang, H. Liu, Effects of different types of biochar on the anaerobic digestion of chicken manure. Bioresour. Technol. **275**, 258–265 (2019)

Chapter 17
Effects of CrN/TiN Coatings on Interfacial Contact Resistance of Stainless Steel 410 Bipolar Plates in Fuel Cells

Mohsen Forouzanmehr, Kazem Reza Kashyzadeh, Amirhossein Borjali, Mosayeb Jafarnode, and Mahmoud Chizari

Abstract Challenge on energy resources exists, especially when the fossil resources are limited. Fuel cells, as an alternative replacement, can be used. Fuel cells with coated bipolar plates are the interest of this paper. Current research is concerned with the effects of CrN/TiN coatings on interfacial contact resistance (ICR). Stainless steel 410 was selected as a base metal, and the coating process was performed using chromium nitride and titanium nitride by cathodic arc evaporation method. It was found that the surface roughness and ICR values of CrN-coated sample are lower than the TiN-coated sample. The concluded that the CrN layer could be replaced with the TiN layer for better performance of bipolar plates.

Keywords Bipolar plate · Titanium nitride coating · Chromium nitride coating · Interfacial contact resistance · Surface roughness

M. Forouzanmehr
Division of Solid Mechanics, School of Science and Engineering, Sharif University of Technology, International Campus, Kish Island, Iran

K. Reza Kashyzadeh (✉)
Department of Mechanical and Instrumental Engineering, Academy of Engineering, Peoples' Friendship University of Russia (RUDN University), Moscow, Russian Federation
e-mail: reza-kashi-zade-ka@rudn.ru

A. Borjali
School of Mechanical Engineering, Sharif University of Technology, Tehran, Iran

M. Jafarnode
School of Mechanical Engineering, Islamic Azad University, South Tehran Branch, Iran

M. Chizari
School of Engineering and Computer Science, University of Hertfordshire, Hatfield, UK

© The Author(s) 2021
I. Mporas et al. (eds.), *Energy and Sustainable Futures*, Springer Proceedings in Energy,
https://doi.org/10.1007/978-3-030-63916-7_17

17.1   Introduction

One of the systems that produce clean energy is the fuel cell. A fuel cell can converts the chemical energy of a fuel directly into electrical energy. In contrast to batteries, fuel cells can generate electricity without the need for recharging until the fuel is fed to the cell. The general reactants of fuel cells are hydrogen and oxygen. Fuel cells as a new generation of clean energy producers have more beneficiaries; including higher performance than other traditional energy sources, simpler design, utilization of hydrogen as a reactor (without environmental pollution), its widespread applications on a small and large scale can be concluded that the designing and optimization of these systems are significant [1]. In this regard, bipolar plates are one of the most significant segments of a fuel cell, which are responsible for separating cells, connecting the cathode to the anode from one cell to another with appropriate electrical conductivity, and feeding reactive gases through gas channels, etc. Accordingly, bipolar plates must have unique electrical, material, and mechanical properties (e.g., high electrical conductivity, high mechanical strength, high corrosion resistance, low surface contact resistance), non-permeability of gas, and low cost [2]. Hence, stainless steel is one of the appropriate choices for making bipolar plates [3].

A considerable improvement on Interfacial Contact Resistance (ICR) of 316L steel bipolar plates with TiN coating has been reported [4]. The published results indicated that the TiN-coated samples had better corrosion resistance compared to non-coated samples. Moreover, TiN and ZrN coatings have been deposited on 316L stainless steel by magnetron sputtering to improve material and electrical properties [5]. The main aim of this research is to study the effects of CrN and TiN coatings on the efficiency of stainless-steel 410 bipolar plates which is our novelty in this study. To achieve this purpose, the ICR and surface roughness of coated samples was compared to that of non-coated samples.

17.2   Material and Specimens

In this research, stainless steel 410 with a thickness of 3 mm was selected as the base material for the bipolar plates. The chemical composition of 410 SS, which was used in this survey was compared with the standard case in Table 17.1. Accordingly, the mechanical properties can be considered as the AISI 410 SS.

Table 17.1   The chemical composition of 410 SS

Sample	Weight percentage of elements (max)						
	C	Mn	Si	P	S	Cr	Ni
410 SS samples	0.13	0.95	1.00	0.03	0.03	13.0	0.70
410 SS [6]	0.15	1.00	1.00	0.04	0.03	13.5	0.75

Table 17.2 Settings for the coating process used in this study

Parameter	Unit	Value
Target metal diameter	mm	80
Current	A	1000
Voltage	V	40
Distance between the specimen and arc head	mm	300
Initial vacuum before coating process	mbar	10^{-6}
Vacuum during coating process	mbar	10^{-3}

To fabricate samples, steel plates were quenched at 1050 °C in an austenitic salt furnace. Furthermore, the tempering operation was performed at 120 °C for 10 min to stress-relief the sheet. Samples with the dimensions of 10*10 mm^2 were cut by utilizing automatic wire cutting machine. At the final stage, a magnetic grinding machine was employed to smooth the surface. The CrN/TiN coatings were deposited by the PVD technique with cathodic-arc evaporation technology and employing one cathode arc head. The settings for this purpose were as described in Table 17.2. Prior to the coating process, the samples were under argon ion bombing for 30 min utilizing level 5 argon gas at a pressure of 10^{-2} mbar and -1000 V bias. Also, the temperature of samples and bias voltage during the coating process were 250 °C and -80 V, respectively. Furthermore, the rotational speed of the samples around the central axis of the chamber and its central axis was 5 and 30 rpm, respectively.

17.3   Experimental Procedure

The Three-Dimensional Atomic Force Microscopy (3D AFM), DME-(DualscopeTM C-26), was used to characterize the morphology of the roughness accurately. Surface roughness determination procedure was obtained using the Lainović study [7]. After that, the interfacial contact resistance between the bipolar plates and the Gas Diffusion Layer (GDL) was measured using Lutron MO-2013 milliohm meter with an accuracy of 1 µΩ. In this test, a gas diffusion layer (Toray TGP-H-060) with a thickness of 190 µm was used to simulate the fuel cell environment. The samples were put between two GDLs and then between two copper plates. Figure 17.1 presents the mechanical setup which is used to apply force (35–150 N cm^{-2}) to this system. Also, each GDL layer was used just once. To ensure the reliability, each experiment was repeated three times, and the average of the surface roughness and ICR were reported. The coating process was applied to one side of the targets. For uncoated samples, the obtained values are divided by 2.

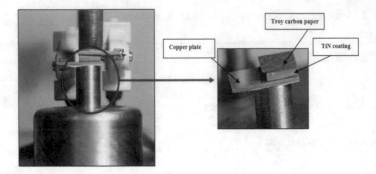

Fig. 17.1 Mechanical setup for applying pressure and measurement of ICR

17.4   Result and Discussion

The 3D AFM surface images for both coated and uncoated specimens are demonstrated in Fig. 17.2. The measured data are reported in Table 17.3 that R_a is the average of surfaces roughness. The results proved that the non-coated samples, TiN-coated samples, and CrN-coated samples have the highest roughness, respectively.

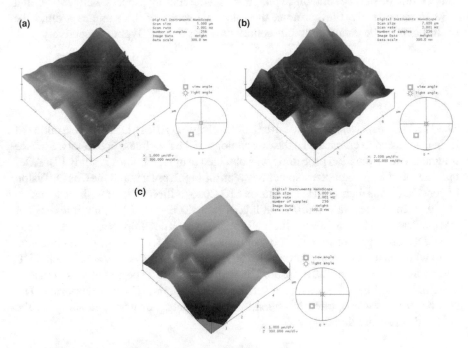

Fig. 17.2 Three-dimensional AFM images for different samples, including **a** CrN-coated sample (PITCH 30°), **b** TiN-coated sample (PITCH 30°), and **c** non-coated sample (PITCH 30°)

Table 17.3 The values of R_a and ICR at 150 N/cm^2

Sample	R_a(nm)	ICR (mΩ cm^2)
CrN-coated	182 ± 11	8.22 ± 0.99
TiN-coated	223 ± 18	9.14 ±1.07
Non-coated	323 ± 28	10.76/2 = 5.38 ±0.12

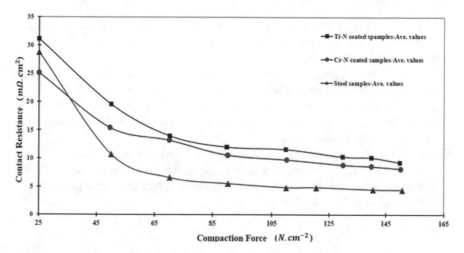

Fig. 17.3 The ICR changes in terms of applied pressure for both coated and non-coated samples

Figure 17.3 depicts the ICR changes in terms of pressure for both coated and non-coated samples. From this figure, it is clear that for all cases, the ICR values decrease by raising the applied pressure. It is because of an increase in the effective contact surface. In addition, the ICR of both coated samples (CrN and TiN-coated) is higher than that of the non-coated sample (Table 17.3). USA Department Energy is recommended ICR values of less than 10 mΩ cm^2 at the pressure of 150 N/cm^2 [7]. Based on statistical analysis, although the average ICR value of the TiN-coated group was higher than CrN-coated group, their differences were insignificant, with 95% of confidence interval (P-Value = 0.336). Moreover, the differences between the ICR values of the Non-coated material group and the TiN-coated group were significant (P-Value = 0.004). Also, the differences between the ICR values of the Non-coated material group and the CrN-coated group were significant, with 95% of confidence interval (P-Value = 0.008). The trend of ICR is a function of surface roughness for coating samples; it means that the CrN-coated bipolar plates have lower ICR values than TiN-coated. The ICR of the CrN-coated sample in the present study is almost 50% of ICR of the CrN-coated steel obtained by PF [8] and PVD methods. Also, the ICR of TiN-coated steel in the present study is approximately 33% of the ICR of TiN-coated steel by EBPVD method.

17.5 Conclusion

In the present research, stainless steel 410 was considered as a base metal for the bipolar plates, and chromium nitride and titanium nitride solid films with the same thickness were deposited on it by the CAE method. The highlight conclusion can be mentioned below:

- Based on the statistical analysis, the ICR value differences of coated samples were significant compared to uncoated samples. It showed that the coated samples increased the ICR values.
- The 3D AFM images, and surface morphology showed that the CrN-coated samples had the minimum values of averaged surface roughness.
- The differences between ICR values of CrN and TiN coated were insignificant, but the average values of ICR and surface roughness of CrN-Coated were lower than that of TiN-Coated group.
- Eventually, it is evident that the coating can improve the corrosion resistance of metals; the lowest values of ICR is required for the efficiency of a fuel cell. Therefore, it can be claimed that the CrN-coated layer is more efficient than a non-coated or TiN-coated layer.

References

1. P.P. Edwards, V.L. Kuznetsov, W.I. David, N.P. Brandon, Hydrogen and fuel cells: towards a sustainable energy future. Energy Policy **36**(12), 4356–4362 (2008)
2. Y. Song, C. Zhang, C.Y. Ling, M. Han, R.Y. Yong, D. Sun, J. Chen, Review on current research of materials, fabrication and application for bipolar plate in proton exchange membrane fuel cell. Int. J. Hydrog. Energy (2019)
3. K. Feng, Y. Shen, H. Sun, D. Liu, Q. An, X. Cai, P.K. Chu, Conductive amorphous carbon-coated 316L stainless steel as bipolar plates in polymer electrolyte membrane fuel cells. Int. J. Hydrogen Energy **34**(16), 6771–6777 (2009)
4. E. Cho, U.S. Jeon, S.A. Hong, I.H. Oh, S.G. Kang, Performance of a 1 kW-class PEMFC stack using TiN-coated 316 stainless steel bipolar plates. J. Power Sources **142**(1–2), 177–183 (2005)
5. P. Yi, L. Zhu, C. Dong, K. Xiao, Corrosion and interfacial contact resistance of 316L stainless steel coated with magnetron sputtered ZrN and TiN in the simulated cathodic environment of a proton-exchange membrane fuel cell. Surf. Coat. Technol. **363**, 198–202 (2019)
6. A.T. Krawczynska, W. Chrominski, E. Ura-Binczyk, M. Kulczyk, M. Lewandowska, Mechanical properties and corrosion resistance of ultrafine grained austenitic stainless steel processed by hydrostatic extrusion. Mater. Des. **136**, 34–44 (2017)
7. T. Lainović, M. Vilotić, L. Blažić, D. Kakaš, D. Marković, A. Ivanišević, Determination of surface roughness and topography of dental resin-based nanocomposites using AFM analysis. Bosn. J. Basic Med. Sci. **13**, 34 (2013)
8. S.J. Lee, C.H. Huang, Y.P. Chen, Investigation of PVD coating on corrosion resistance of metallic bipolar plates in PEM fuel cell. J. Mater. Process. Technol. **140**(1–3), 688–693 (2003)

Chapter 18
Moisture Stable Soot Coated Methylammonium Lead Iodide Perovskite Photoelectrodes for Hydrogen Production in Water

Udit Tiwari and Sahab Dass

Abstract Metal halide perovskites have triggered a quantum leap in the photovoltaic technology marked by a humongous improvement in the device performance in a matter of just a few years. Despite their promising optoelectronic properties, their use in the photovoltaic sector remains restricted due to their inherent instability towards moisture. Here, we report a simple, cost-effective and highly efficient protection strategy that enables their use as photoelectrodes for photoelectrochemical hydrogen production while being immersed in water. A uniform coating of candle soot and silica is developed as an efficient hydrophobic coating that protects the perovskite from water while allowing the photogenerated electrons to reach the counter electrode. We achieve remarkable stability with photocurrent density above 1.5 mA cm^{-2} at 1 V versus saturated calomel electrode (SCE) for ~1 h under constant illumination. These results indicate an efficient route for the development of stable perovskite photoelectrodes for solar water splitting.

Keywords Photoelectrochemical water splitting · Halide perovskites · Hydrogen · TiO$_2$ · CH$_3$NH$_3$PbI$_3$

18.1 Introduction

The world has entered into an era where climate change has become an unfortunate reality [1]. Any further increase in the CO$_2$ levels might push the Earth's climate over a tipping point from which recovery would become impossible [2]. The unprecedented pace at which global warming is damaging the earth is evident from the warming of the oceans and the accelerated melting of the polar ice caps all of which point to the increasing level of the greenhouse gases [3]. There is thus an urgent need for the world

U. Tiwari (✉)
School of Physical Sciences and Computing, University of Central Lancashire, Preston PR1 2HE, UK
e-mail: UTiwari1@uclan.ac.uk

S. Dass
Department of Chemistry, Dayalbagh Educational Institute, Agra 282005, India

© The Author(s) 2021
I. Mporas et al. (eds.), *Energy and Sustainable Futures*, Springer Proceedings in Energy,
https://doi.org/10.1007/978-3-030-63916-7_18

to shift from the conventional carbon-based fuels to a renewable energy source to tackle global warming while also fulfilling the global energy demands. In this regard, hydrogen can be seen as the best alternative energy source as it is absolutely clean (C/H ratio is 0), freely available (most abundant element), has a very high energy per unit mass (1 kg of hydrogen provides the same power as 3.78 L of gasoline) and has a high energy storage capacity per mole [4]. Currently, hydrogen production mainly relies on methane steam reforming and electrolysis both of which cannot be considered environment-friendly and thus defeats the purpose of producing hydrogen [5]. Amongst renewable methods of hydrogen production, the photoelectrochemical splitting of water is considered to be the most promising method which uses semi-conductors submerged in aqueous electrolyte as photo-absorbers for sunlight driven splitting of water [6]. However, the development of inexpensive and moisture stable visible light-absorbing materials is still a challenging task [7]. Hybrid Organometal halide perovskites offer great promise as photoelectrode materials thanks to their low cost, easy processability, high efficiency [8], tunable band-gap [9], ambipolar charge transport [10], long carrier lifetimes [11], and long charge diffusion lengths [12]. The immense potential of these materials is evident from the rapid improvement in their power conversion efficiencies from over 3.8% in 2009 to values reaching over 23% in 2018 making it the fastest growing technology in the history of photovoltaics [13]. Despite the tremendous promise shown by these materials, their wide-scale commercialisation is restricted owing to one major drawback; they are inherently unstable in water. The lattice of methylammonium lead iodide perovskite degrades even with the slightest exposure to moisture, followed by the decomposition of the material into PbI_2 [14]. Several attempts have been made to improve the stability of halide perovskites towards moisture, including protective surface coatings, use of more hydrophobic alkyl ammonium salts, atomic layer deposition or through 2D/3D hybrid structures [15]. These protection strategies are not only expensive but also fail to impart long term stability to the perovskite photoelectrodes and therefore are difficult to upscale.

Here we report a simple, cost-effective and efficient protection strategy to enable the perovskite photoelectrodes to produce hydrogen while being submerged in an aqueous electrolyte. A thin layer of silica over a uniform layer of candle soot is used to effectively protect the moisture-sensitive $CH_3NH_3PbI_3$ (referred to as $MAPbI_3$) perovskite layer from degradation in water. The combination of carbon and silica served as an effective hydrophobic layer shielding the photoelectrodes from water. The m-TiO_2|$MAPbI_3$|C|SiO_2 photoanodes operate in water, exhibiting photocurrents of over 1.8 mA/cm^2 at an applied bias of 1 V in an alkaline solution (pH 13). Even more impressively, these photoelectrodes exhibit remarkable stability showing no considerable change in the current for over 1 h under constant illumination.

18.2 Experimental

18.2.1 Synthesis of Methylammonium Lead Iodide

30 mL Hydroiodic acid (57 wt.% in water, Aldrich) was added dropwise to 27.86 mL methylamine (40% in methanol, Aldrich) in a 250 mL round-bottomed flask and was subjected to constant magnetic stirring at 0 °C for 2 h. Crystallization was done by evaporating the solvents at 50 °C for 1 h. The obtained powder was purified by washing three times with diethyl ether, followed by drying at 60 °C in a vacuum oven for 24 h. The product, methylammonium iodide was saved at room temperature in a desiccator [16].

18.2.2 Solar Cell Fabrication

A compact layer of TiO_2 as an electron transport layer (ETM) was deposited onto the FTO substrates by spin coating a solution containing 0.65 ml of Ti (IV) isopropoxide (Sigma Aldrich, 97%), 0.38 ml of Acetylacetone (Sigma Aldrich, reagent grade) and 5 ml of Ethanol at 3000 rpm for 60 s. The substrates were then calcined at 500 °C for 30 min to obtain the dense TiO_2 layer. The perovskite precursor solution was prepared by mixing $CH_3NH_3PbI_3$ and PbI_2 (1:1 mol ratio) in γ-butyrolactone at 60 °C for 12 h with constant magnetic stirring. The resulting solution was coated onto the FTO/TiO_2 substrate at 2,000 rpm for 60 s then at 3,000 rpm for 60 s, and dried on a hot plate at 100 °C for 2 min [16]. A uniform layer of soot was immediately deposited onto the FTO/TiO_2/$MAPbI_3$ substrate with the help of a burning candle, followed by the spin-coating of a uniform layer of SiO_2 (0.1 g SiO_2 in 2 mL γ-butyrolactone) over it. These FTO/TiO_2/$MAPbI_3$/C/SiO_2 substrates (see Fig. 18.1) were developed into photoelectrodes by establishing ohmic electrical contacts using

Fig. 18.1 3D configuration of soot coated perovskite device

FTO TiO2 MAPbI3 C/SiO2

silver paste and copper wire from the uncoated area of the conductive substrates. The area of contact was then covered with non-transparent and non-conducting epoxy-resin (Hysol, Singapore).

18.3 Results and Discussion

18.3.1 Analysis of the Crystal Structure

Figure 18.2 shows the XRD pattern of the soot coated perovskite thin film. The peaks at 14.08°, 28.36°, 31.76°, and 43.08° correspond to the tetragonal/β phase of MAPbI$_3$ oriented along the (110), (220), (310,) and (330) planes respectively (JCPDS 00-021-1276). The peaks at 25.304°, 37.793° and 48.037° correspond to the (101), (004), and (200) planes respectively of the tetragonal phase of TiO$_2$ (JCPDS 65-5714). The peaks at 2θ values of 15.194°, 16.966° and 26.544° correspond to the (110), (003) and (300) planes respectively of carbon (JCPDS 50-0927), while those at 10.892° and 40.265° correspond to the (101) and (005) planes respectively of SiO$_2$ (JCPDS 771414). All other peaks correspond to the FTO substrate. The XRD pattern reveals that both MAPbI$_3$ and TiO$_2$ are obtained in a highly crystalline state while also indicating the presence of carbon and SiO$_2$. Moreover, the coating of soot over the perovskite layer does not alter its crystalline structure suggesting that carbon did not incorporate into the crystal lattice of the perovskite.

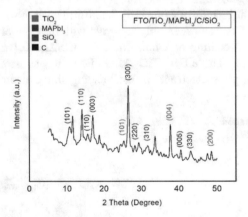

Fig. 18.2 XRD pattern of the FTO/TiO$_2$/MAPbI$_3$/C/SiO$_2$ photoelectrode

18.3.2 Optical Measurements

The optical absorbance measurements were made using a UV-Visible Spectropho-
tometer (Shimadzu, UV-2450, Japan) in the wavelength range of 800 to 200 nm
using a UV quartz sample cell with a transmission range of 190–2500 nm. The
UV-Visible spectra (Fig. 18.3a) compares the absorption characteristics of the
pure perovskite (FTO/MAPbI₃) and the soot coated perovskite photoelectrode
(FTO/TiO₂/MAPbI₃/C/SiO₂). The spectra show that the incorporation of TiO₂ and
carbon as electron transport layer (ETL) and hole transport layer (HTL) respectively
results in the shifting of the absorption towards the ultraviolet region. The band-gap
energies were calculated by the absorbance data using the following equation:

$$\alpha h\upsilon = A\left(h\upsilon - E_g\right)^m \tag{18.1}$$

where α is the absorption coefficient, $h\upsilon$ is the photon energy in eV, and E_g is the
band-gap energy in eV. A is a constant related to the effective mass of the electrons
and holes and m is equal to 0.5 for allowed direct transition and 2 for an allowed
indirect transition. Plots between $(\alpha h\upsilon)^2$ and $h\upsilon$ for the pristine perovskite and the
soot coated perovskite are shown in Fig. 18.3b, c respectively. The linear nature of
the graph suggests that the sample behaves as a direct band-gap material. The band-
gap for the pristine perovskite sample comes out to be 2.05 eV close to the reported
value of 1.56 eV [7] while for the soot coated perovskite, it comes out to be 2.49 eV.
The high band gap SiO₂ layer may have resulted in the overall increase in the device
band gap.

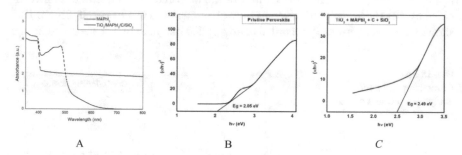

Fig. 18.3 a Absorption spectra of the FTO/MAPbI₃ and FTO/TiO₂/MAPbI₃/C/SiO₂ thin films.
b Tauc plot for the pristine perovskite sample (FTO/MAPbI₃). c Tauc plot for the soot coated
perovskite sample

Fig. 18.4 Photocurrent
density voltage
characteristics of the soot
coated perovskite
photoelectrode in 0.1 M
NaOH electrolyte

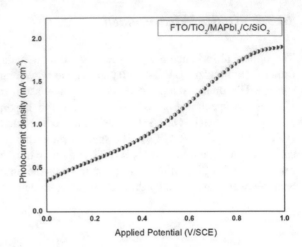

18.3.3 Photoelectrochemical Measurements

The photosensitivity of the prepared electrode is analysed by investigating its current versus voltage characteristics in a three-electrode quartz cell under light and dark conditions. Figure 18.4 shows the photocurrent density of the photoelectrode as a function of the electrode potential. The $FTO/TiO_2/MAPbI_3/C/SiO_2$ combination yielded a photocurrent density of 1.8 mA cm^{-2} at an applied bias of 1 VSCE. The excellent photo-response of the soot coated perovskite photoelectrode suggests that the hydrophobic soot/SiO$_2$ coating did not interfere with the photoelectrochemical response of the perovskite. The stability of the prepared photoelectrodes is investigated by measuring the current versus time characteristics in a PEC cell. Figure 18.5 shows Chronoamperometry recorded at an applied potential of 1 V versus SCE in a

Fig. 18.5
Chronoamperometric
analysis of the soot coated
photoelectrode

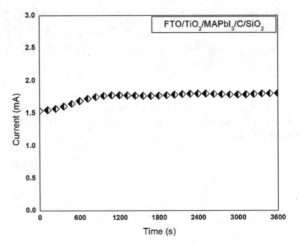

0.1 M aqueous NaOH solution. The soot coated photoelectrodes showed remarkable stability with currents remaining above 1.5 mA cm^{-2} for ~1 h under constant illumination. This represents a first example of halide perovskite photoelectrode, protected with an inexpensive protection layer.

18.4   Conclusion

This work focusses on solving the long-standing moisture instability issue of halide perovskites through a simple, efficient and inexpensive protection route. The FTO/TiO$_2$/MAPbI$_3$/C/SiO$_2$ photoelectrodes exhibited the potential to be employed as stable electrodes for water splitting in PEC hydrogen production, showing outstanding stability for ~1 h in aqueous electrolyte. The perfect energy band alignment permitted carbon to be an efficient and directional hole extraction layer. While the excited electron moves from the conduction band (CB) at -3.86 eV of the MAPbI$_3$ to the CB of TiO$_2$ (-4.00 eV), the hole in the valence band (VB) at -5.43 eV is effectively extracted to the carbon layer (work function 5.00 eV). While the candle soot acted as an effective HTL due to its intimate contact with the perovskite layer, the SiO$_2$ offered a more compact seal that protected the perovskite from the liquid electrolyte. The enormous stability coupled with the high photocurrent density makes these soot coated photoelectrodes an attractive device for application in a tandem photoelectrochemical cell for solar water splitting. Further research into the optimization of the electron transport layer and the use of more efficient perovskites can improve both stability and efficiency.

Acknowledgements This research has been supported by the Department of Chemistry, Dayalbagh Educational Institute, Agra, India and MNRE, Govt. of India project No. 103/241/2015-NT. We are thankful to the Department for providing the experimental and characterization facilities. We also thank Dr Karen Syres (University of Central Lancashire) for many valuable suggestions.

References

1. UN Framework Convention on Climate Change Secretariat, Paris Agreement FCCC/CP/2015/L.9/Rev.1 (2015)
2. S. Tapio, M.K. Colleen, G.P. Kyle, Possible climate transitions from breakup of stratocumulus decks under greenhouse warming. Nat. Geosci. **12**, 163–167 (2019)
3. L. Sydney et al., World ocean heat content and thermosteric sea level change (0–2000 m), 1955–2010. Geophys. Res. Lett. **39**, 10 (2012)
4. B. Tadeusz, T.N. Janusz, R. Mieczyslaw, S. Charles, Photo-electrochemical hydrogen generation from water using solar energy. Materials-related aspects. Int. J. Hydrog. Energy **27**, 991–1022 (2001)
5. N.Z. Muradov, T.N. Veziroğlu, From hydrocarbon to hydrogen-carbon to hydrogen economy. Int. J. Hydrog. Energy **30**, 225–237 (2005)

6. B. Dowon, S. Brian, C.K.V. Peter, H. Ole, C. Ib, Strategies for stable water splitting: via protected photoelectrodes. Chem. Soc. Rev. **46**, 1933–1954 (2017)
7. J.J. Nam et al., Compositional engineering of perovskite materials for high performance solar cells. Nature **517**, 476–480 (2015)
8. H.N. Jun, H. Sang, H.H. Jin, N.M. Tarak, I.S. Sang, Chemical management for colorful, efficient, and stable inorganic–organic hybrid nanostructured solar cells. Nano Lett. **13**, 1764–1769 (2013)
9. B. Sai et al., High-performance planar heterojunction perovskite solar cells: preserving long charge carrier diffusion lengths and interfacial engineering. Nano Res. **7**, 1749–1758 (2014)
10. W. Christian, E. Giles, B.J. Michael, J.S. Henry, M.H. Laura, High charge carrier mobilities and lifetimes in organolead trihalide perovskites. Adv. Mater. **26**, 1584–1589 (2014)
11. S. Dong et al., Low trap-state density and long carrier diffusion in organolead trihalide perovskite single crystals. Science **347**, 519–522 (2015)
12. NREL's 'Best Research-Cell Efficiencies'. https://www.nrel.gov/pv/assets/pdfs/pv-efficiencies-07-17-2018.pdf. Accessed 14 December 2018
13. P. Bertrand, P. Byung-Wook, L. Rebecka, O. Johan, A. Sareh, M.J.J. Erik, R. Hakan, Chemical and electronic structure characterization of lead halide perovskites and stability behavior under different exposures—a photoelectron spectroscopy investigation. Chem. Mater. **27**, 1720–1731 (2015)
14. K. Min, G.M. Silvia, S. Roberto, P. Annamaria, Enhanced solar cells stability by hygroscopic polymer passivation of metal halide perovskite thin film. Energy Environ. Sci. **11**, 2609–2619 (2018)
15. P. Isabella, E. Salvador, C. Petra, Tetrabutylammonium cations for moisture-resistant and semitransparent perovskite solar cells. J. Mater. Chem. A **5**, 22325–22333 (2017)
16. Z. Masahito, S. Atsushi, A. Tsuyoshi, O. Takeo, Fabrication and characterization of $TiO_2/CH_3NH_3PbI_3$-based photovoltaic devices. Chem. Lett. **43**(6), 916–918 (2014)

Chapter 19
Low-Speed Aerodynamic Analysis Using Four Different Turbulent Models of Solver of a Wind Turbine Shroud

M. M. Siewe Ngouani, Yong Kang Chen, R. Day, and O. David-West

Abstract This study presents the effect of four different turbulent models of solver on the aerodynamic analysis of a shroud at wind speed below 6 m/s. The converting shroud uses a combination of a cylindrical case and an inverted circular wing base which captures the wind from a 360° direction. The CFD models used are: the SST (Menter) k-ω model, the Reynolds Stress Transport (RST) model, the Improved Delay Detached Eddies Simulation model (IDDES) SST k-ω model and the Large Eddies Simulation Wall Adaptive model. It was found that all models have predicted a convergent surface pressure. The RST, the IDDES and the WALE LES are the only models which have well described regions of pressure gradient. They have all predicted a pressure difference between the planes (1–5) which shows a movement of the air from the lower plane 1 (inlet) to the higher plane 5 (outlet). The RST and IDDES have predicted better vorticities on the plane 1 (inlet). It was also found that the model RST, IDDES, and WALE LES have captured properly the area of turbulences across the internal region of the case. All models have predicted the point of flow separation. They have also revealed that the IDDES and the WALE LES can capture and model the wake eddies at different planes. Thus, they are the most appropriate for such simulation although demanding in computational power. The movement of air predicted by almost all models could be used to drive a turbine.

Keywords Low-speed aerodynamic · Wind turbine shroud · Turbulence models · Coanda effect · Flow direction change

19.1 Introduction

The recent problem of global warming and the concern for a sustainable energy resource for a better world have led to the development of wind turbines technologies to harvest the power of wind anyhow [1]. It has been noted that amongst the categories of wind generators, small wind turbines have been more and more popular for their

M. M. S. Ngouani (✉) · Y. K. Chen · R. Day · O. David-West
School of Engineering and Computer Science, University of Hertfordshire, Hatfield AL10 9AB, Herts, UK
e-mail: m.ngouani-siewe3@herts.ac.uk

© The Author(s) 2021
I. Mporas et al. (eds.), *Energy and Sustainable Futures*, Springer Proceedings in Energy,
https://doi.org/10.1007/978-3-030-63916-7_19

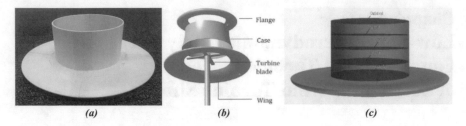

Fig. 19.1 Shroud view: (**a**) Printed wind turbine case; (**b**) Overall Schematics of the wind turbine; (**c**) Planes from which data has been investigated: from bottom to top (plane 1, 2, 3, 4, 5)

excellent adaptability to the urban area in terms of noise pollution and visual impact [2]. However, these small wind turbines have only been operated at wind speeds greater than 6 m/s [3]. In addition they have been found to produce still important turbulences, thus noise [4]. The purpose of this study is to identify the best turbulent model that would properly capture and characterise the nature of air flow inside and around the shroud. Thus, this paper presents a comparative aerodynamic analysis of the performance of a converting shroud to be used in a wind turbine system working at wind speed below 6 m/s using the software package Star-CCM turbulence models. The turbulence models investigated are notably: the SST (Menter) k-ω model, the Reynolds Stress Transport model, the Improved Delay Detached Eddies Simulation model (IDDES) SST k-ω model, the Large Eddies Simulation Wall Adaptive model (Fig. 19.1).

19.2 Experimental

19.2.1 K-ω Turbulent Model

It comprises modifications for low Reynolds number effects, compressibility and shear flow spreading compare to the realizable k-ε. It is characterized by the turbulent Kinetic energy and the frequency $\omega = k/\varepsilon$, where ε is the rate of dissipation of k. The SST model has been widely used in the aerospace industry, where viscous flows are typically well resolved and turbulence models are generally applied throughout the boundary layer. One advantage of k-ω model is its improved performance for boundary layers under adverse pressure gradients.

19.2.2 The Reynolds Stress Turbulent Model

The RST model has the greatest potential to accuracy. However, its results are still compromised by model assumptions and the use of the RST model does not justify the extra computational effort for simple flows. They solve transport equations for all components of the specific Reynolds stress tensor. They can account for anisotropy effects due to strong swirling motion, streamline curvature, rapid changes in strain rate and secondary flows in ducts.

19.2.3   Detached Eddy Simulation: DES (IDDES SST k-ω Turbulence Model)

The DES-SST method is a unified LES/RANS hybrid which separates the domain into a near-wall region where RANS equations are solved and an outer region where LES equations are solved. This method is very dependant of the properties of the grid. The distinction between the two sets of equations is only done by the source term in the transport equation for a turbulence quantity. The idea of DES can however be extended to any specific turbulence model and a combination with the SST model exists which is evaluable on Star-CCM+.

According to Shur et al., the IDDES model provides a more flexible and convenient scale-resolving simulation model for high Reynolds number flows. Due to the fact that IDDES combines DES and wall-modelled LES, this new model helps in solving the grid-induced separation as it increases the modelled stress contribution across the interface.

19.2.4   Large Eddy Simulation (LES WALE)

Another turbulent model used in this research is the Large-Eddy Simulation (LES). However, this model describes high Reynolds Number time-evolving, three-dimensional turbulence. LES methods resolve the largest turbulent scales within a flow, and filter the smaller scales (dependent on mesh resolution) using various sub-grid scale models. The use of this approach requires careful application of the model, and significant computational resource. WALE (Wall-Adapting Local Eddy-viscosity) chose in this study provides zero eddy viscosity when dealing with laminar flow which is important for transition.

19.2.5   Geometry and Mesh Generation

The near wall was set to low y+. The number of prism layer used was 20 and the overall boundary layer was resolved. The mesh model was the unstructured polyhedral model. This achieved a number of cells of approximately 13.4 million (Figs. 19.2 and 19.3).

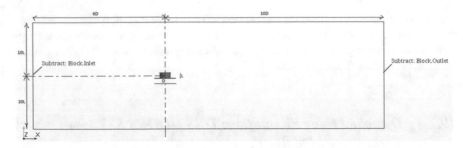

Fig. 19.2 Schematic view of the size of the domain of the CFD simulation within Star-CCM+

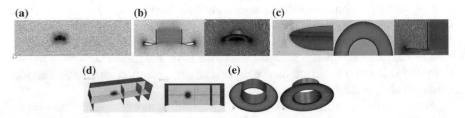

Fig. 19.3 Mesh generation around the shroud: **a** mesh structure within the domain; **b** section cut view of the mesh within the shroud; **c** prism layers distribution near the shroud; **d** mesh in bottom view of the shroud; **e** surface mesh on the shroud

19.3   Results and Discussion

Figure 19.4 shows the internal distribution of velocity and areas of vorticities. It can be seen that; the models RST, IDDES, and WALE LES capture properly the area of turbulences across the internal region of the case. However, only the LES WALE presents area of turbulences at inlet. In addition, all models, clearly show a region of low velocity and a region of high velocity. The latter region represents about a quarter of the whole cross section from planes 2 to 5.

The external velocities and wake distribution in the Fig. 19.5 reveals that the IDDES and the WALE LES can capture and model the wake eddies across the different planes. The SST k-ω does not capture any vorticity at all.

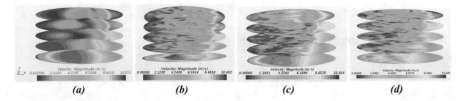

Fig. 19.4 Internal velocity scalar on the planes 1, 2, 3, 4, 5: **a** SST (MENTER) k-ω model, **b** (RST) Reynolds stress turbulence, **c** IDDES SST k-ω turbulence model, **d** LES WALE turbulence model

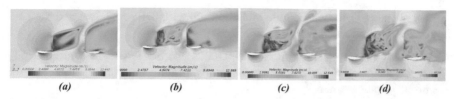

Fig. 19.5 External velocity scalar on coordinate planc (0, X, Y): **a** SST (MENTER) k-ω model, **b** (RST) Reynolds stress turbulence, **c** IDDES SST k-ω turbulence model, **d** LES WALE turbulence model

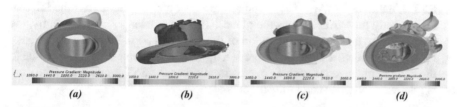

Fig. 19.6 Pressure gradient: **a** SST (MENTER) k-ω; **b** RST model; **c** IDDES SST k-ω; **d** LES WALE

Table 19.1 CFD parameters at 6 m/s

	SST k-ω	IDDES SST k-ω	RST	LES WALE
F (N)	−3.45	−2.56	−2.76	−2.8234
D (N)	2.27	2.29	2.145	2.045
Mass flow inlet plane 1 (kg/s)	0.236	0.242	0.231	0.235
Mass flow outlet plane 5 (kg/s)	0.096	0.13	0.10	0.12

The pressure gradient is an excellent parameter in determining the region of flow separation on the wing-surface or within the region. Hence, it can be observed in Fig. 19.6 that all models have predicted that the front top and bottom parts of the wing are subjected to separation and attachment. The reattachment is shown by the darker blue regions on the surface of the shroud.

Table 19.1 shows a summary of the different aerodynamic parameters calculated in the study. It can be seen that there is a difference in the mass flow rate between the inlet and the outlet. Therefore, there is a speed variation between those 2 planes.

19.4   Conclusion

The CFD investigation of the shroud has involved the use of 4 turbulent models such as: the SST (menter) k-ω model, the RST Reynolds Stress Turbulent model, the improved Delay Detached Eddies simulation model IDDES SST k-ω and the LES large Eddies simulations WALE.

- The study has revealed the RST, the IDDES and the WALE LES are the only models which described external pressure gradient regions well and therefore the shroud turbulent wake.
- Internally, all models predicted a pressure difference between the planes 1 and 5 which shows a movement of the air from the lower plane 1 (inlet) to the higher plane 5 (outlet). This motion could be used to drive a turbine.
- The RST and IDDES predicted better vorticities on the plane 1 (inlet). Although RST, IDDES, and WALE LES captured areas of turbulences across the internal region, only the WALE shows the plane 1 (inlet) turbulences. Subsequently, the study showed that the internal region of the shroud is partially highly turbulent.
- Finally, all models showed that there is a downward lift which is produced due to the wing being inverted with an overestimation on the SST k-ω model. The drag is relatively the same across the models.

Acknowledgements This project has been supported by the University of Hertfordshire Cluster team and the technician team.

References

1. T. Burton, D. Sharpe, N. Jenkins, E. Bossanyi, Wind energy handbook. Chichester, UK (2013)
2. J.F. Manwell, J.G. McGowan, A.L. Rogers, *Wind Energy Explained-Theory, Design and Application* (Wiley, Chichester, UK, 2012).
3. K. Pope, I. Dincer, G.F. Naterer, Energy and exergy efficiency comparison of horizontal and vertical axis wind turbines. Renew. Energy **35**(9), 2102–2113 (2010)
4. A.L. Rogers, J.F. Manwell, S. Wright, Wind Tubrine Acoustic Noise: Renewable Energy Research Laboratory, University of Massachusetts at Amherst, USA (2016)

Chapter 20
Design Procedure of a Hybrid Renewable Power Generation System

Seyed Vahid Hosseini, Ali Izadi, Seyed Hossein Madani, Yong Chen, and Mahmoud Chizari

Abstract Electrification of small communities in districted off-grid area remains as a challenge for power generation industries. In the current study, various aspects of design of a standalone renewable power plant are examined and implemented in a case study of a rural area in Cape Town, South Africa. Estimating required electricity based on local demand profile, investment, operability, and maintenance costs of different generation technologies are studied in order to investigate their potential in an off-grid clean energy generation system. Several configurations of hybridization of solar system, wind, and micro gas turbine in combination with a battery are investigated. The Levelized Cost of Electricity (LCOE) and number of days with more than 3 h black out are compared.

Keywords Hybrid power generation · Solar · Wind · Energy · Micro gas · Micro grid · Micro power plant

20.1 Introduction

Renewable systems have shown to have great potential to be employed in remote areas since they do not need expensive structure and complicated grid infrastructure [1]. A possible configuration for renewable generators and storages focusing on a combination of PV, wind, hydrogen fuel cell, battery, and pumped hydro is provided in [2]. While it is important to consider weather condition for the wind and solar generation in order to define optimal micro-grid operation [3].

Current article is prepared to address the first step to provide a procedure to evaluate the performance and economy of a hybrid renewable system with the required

S. V. Hosseini (✉) · Y. Chen · M. Chizari
Univesity of Hertfordshire, Hatfield, UK
e-mail: v.hosseini@herts.ac.uk

A. Izadi · S. H. Madani
Samad-Power Ltd., Milton Keynes, UK

I. Mporas et al. (eds.), *Energy and Sustainable Futures*, Springer Proceedings in Energy,
https://doi.org/10.1007/978-3-030-63916-7_20

battery capacity. A remote area in South Africa is chosen to compare various genera-tion systems. Single source power plant with Wind, Solar PV and Micro Gas Turbine (MGT), as well as dual combination of these sources, are considered. Then it will be possible to optimize the hybridizing which will be reported in a separated article.

20.2 Theory and Methodologies

The required information for the analysis is mainly extracted from three different sources. (1) System Advisor Model (SAM) [4] which is an open source code developed by the National Renewable Energy Laboratory (NREL); (2) Photovoltaic Geographical information system [6] also provided by the European commission to extract solar energy; (3) The Global Atlas [5] provided by International Renewable Energy Agency to extract wind energy data.

For the MGTs (Micro gas turbines), characteristics of Capstone C35, C60 and C200S are considered. The wind energy assumed to be generated with a number of SAM generic 100 kW turbines. For the Solar PV, also, the solar library of SAM was deployed. The case study is demonstrated for a rural area near Cape Town. SAM software is utilized to calculate techno-economic parameters for each renewable energy generators. Both the demand and generation profile for the area is extracted in daily, monthly and annual manner.

20.2.1 Profile Demand

Total annual demand of the selected area is estimated to be around 1500 kWh to supply 2500 people in a rural area using the World Bank [7] framework of South Africa plan. A similar profile demand in the SAM is normalized to meet this annual value. Since the case study is located in the southern hemisphere, more electricity is required from January to March and October to December for cooling purposes.

20.2.2 Solar Power Generation

Solar radiation data for the case study area is extracted from [8] then the Solar PV array is simulated using the SAM to supply the required demand. The monthly clearness index, defined as the fraction of solar radiation at the top of the atmosphere that reaches a particular location on the earth surface, is also considered varying between 0.495 (in August) and 0.586 (in October) with an annual average of 0.543.

20.2.3 Wind Power Generation

The wind speed data for the case study location is obtained from the Global Wind Atlas (GWA). A wind farm with the NREL 100kW turbine, which is a reference turbine in the SAM, is selected to supply the load demand. The shear coefficient and hub height of this wind turbine are 0.14 and 80 m, respectively. The hourly variation of the wind speed in the hub height is considered for the simulation.

20.2.4 Micro Gas Turbine Power Generation

According to market study considering site demands and the commercial micro gas turbine, Capstone C30, C65 and C200S are selected as the available MGT engines. Nominal characteristics of the each MGT is obtained from the Gas Turbine World magazine [9] and repair and maintenance of the MGTs is assumed to be 0.015 $/kWh. The effect of ambient condition on the performance of these engines is evaluated based on their correction curves [10]. Also, based on diesel wholesale price (0.344 $/lit) and the heating values of 45.6 MJ/kg in South Africa, and the efficiency of MGTs, fuel cost is calculated to be 0.10, 0.11 and 0.13 $/kWh for C200S, C65 and C30, respectively.

20.2.5 Battery Pack Sizing

The battery capacity is calculated in a way to cover the intermittency of the renewable sources in 24 h supply. To do so, the difference between demand load and power generation in the past 24 h of each hour for the whole year is calculated and the mode (charging/discharging) is determined. The accumulated of charging/discharging values within a duration defines the size of the battery pack.

20.3 Financial Modelling

To consider both capital and operating expenses (CAPEX and OPEX) for a generation system, 25 years lifetime is considered for the system. Based on available information, the current interest rate and inflation rate in Africa are 16% and 13.5% respectively [11] which results in 2.2% for the annual real interest rate. Moreover, to compare the economic aspect of the systems, it is assumed that the extra generation of the system could not be exported to the grid.

20.4   Results and Discussion

This section investigates several configurations for utilization of the renewable sources for the case study area. At first step single power generator (solar PV or wind turbine or gas turbine) and then, the combination of power plants including two generators (Solar PV and Wind turbine, Solar PV and Gas turbine, wind turbine and gas turbine) with battery packs are simulated. Although the best share could be evaluated with optimization processes, shares of each power sources in this section are assessed in a heuristic manner with the target of obtaining a suitable fit on annual demand profile. Capacity of the power plant is adjusted to supply at least 90% of monthly demand load.

20.4.1   Single Power Generation Plants

A comparison between generations of the three single-source power plants which is plotted beside the monthly demand is shown in Fig. 20.1. As it could be seen, the MGT has the most uniform generation (due to less variation of the ambient condition of the site) and on the contrary, a Solar PV system to provide at least 90% of the demand, needs to be sized equivalent of two times of the annual demand. The worst 24 h of each generation system is plotted in Fig. 20.2. These graphs clearly show that for a PV system the daily difference between generation and demand profile is much higher in comparison with wind and MGT.

Besides the cost of electricity, sustainability of the generation is another important parameter in assessment of a plant. In this order, the number of days in which there is

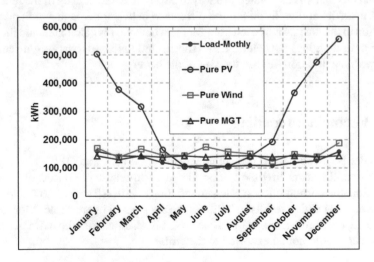

Fig. 20.1 Load versus Generation for single source power generation plants

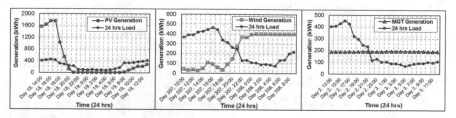

Fig. 20.2 Load versus Generation for worst 24 h

Table 20.1 Comparison between single source power plants

Techno-economic parameters		Solar	Wind	MGT
Technical parameters	Annual generation (kWh)	3,401,435	1,850,304	1,530,858
	Rating capacity (kW)	2680	400	200
	Capacity factor %	14.5	52.8	87.4
	Averaged generation per month (kWh)	279,570	154,192	127,571
	Averaged generation per day (kWh)	9,319	5,069	4,194
	Battery capacity (kWh)	2,680	3239	1425
	Averaged extra generation per day (kWh)	5,189	960	85
	Days with more than 3 h of no electricity access	36	47	0
Economic parameter	CAPEX ($)	2,631,200	1,280,000	225,000
	OPEX ($)	22,770	8,000	163,000
	LCOE (¢/kWh)	18.7	8.9	12.1

no supply for more than 3 h is indicated in Table 20.1. The wind power plant with the lowest value of LOCE has 47 days with more than 3 h completely off in contrary to MGT highest value of LOCE and no day in this condition. These results emphasize that hybridization could provide the possibility to enhance the overall outage of the plant.

20.4.2   Hybrid Power Generation Plants

To demonstrate the advantages of each generation system, results of combination of each two of them are presented in this section. Figure 20.3 shows both the monthly generation and monthly share of these hybrid systems in comparison with monthly demand. While the generation and demand profiles for a day with the worst condition

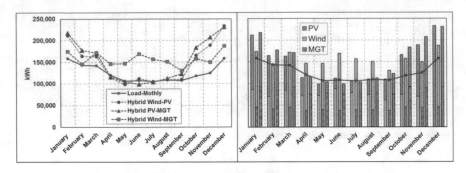

Fig. 20.3 Load versus Generation for hybrid power generation plants

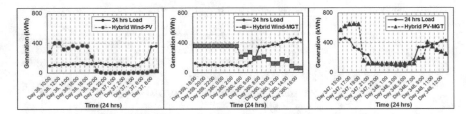

Fig. 20.4 Load versus Generation for worst 24 h of hybrid system

are used to size the battery which is plotted in Fig. 20.4. This condition occurs in January for the Wind-PV and in December for the other two systems.

Table 20.2 summarizes the overall techno-economic parameters of these systems. Hybridization has a great effect not only in reduction of LCOE of PV system but also in decreasing the number of days with more than 3 h of completely without generation. For the PV-Wind system, single PV and single wind, as shown in Table 20.1, there are 36 and 47 days in this condition, respectively; but the hybrid PV-Wind has only 25 days with more than 3 h completely without generation.

20.5 Conclusion

Solar PV which could provide at least 90% of monthly demand, generate twice the annual demand. By combination, the synchronization between generation and demand is enhanced. So, the Levelized Cost of Electricity is reduced. Moreover, to consider the quality of power generation, the number of days which have more than 3 h with completely no electricity is compared between different scenarios that show a noticeable improvement in combined systems. This study believes that optimization techniques considering various parameters are required to find the best solution for each case study.

Table 20.2 Comparison between hybrid power plants

Techno-economic parameters		PV and Wind	Wind and MGT	PV and MGT
Technical parameters	Annual generation (kWh)	1,775,940 PV 48% Wind 52%	1,885,256 Wind 74% MGT 26%	1,845,845 PV 46% MGT 54%
	Rating capacity (kW)	PV: 670 Wind: 200	Wind: 300 MGT: 65	PV: 650 MGT: 2 × 65
	Capacity factor %	23	59	26
	Averaged generation per month (kWh)	147,995	157,105	153,820
	Averaged generation per day (kWh)	4,866	5,165	5,057
	Battery capacity (kWh)	2,057	2,796	1,172
	Averaged extra generation per day (kWh)	756	1,055	947
	Days with more than 3 h of no electricity access	25	0	0
Economic parameter	CAPEX ($)	1,297,000	1,060,000	857,000
	OPEX ($)	9,693	64,000	121,693
	LCOE (¢/kWh)	9.1	11.2	13.7

References

1. N. Izadyar, H.C. Ong, W.T. Chong, K.Y. Leong, Resource assessment of the renewable energy potential for a remote area: a review. Renew. Sustain. Energy Rev. **62**, 908–923 (2016)
2. K.R. Khalilpour, A. Vassallo, A generic framework for distributed multi-generation and multi-storage energy systems. Energy **114**, 798–813 (2016)
3. W.D. Xu, Recent advance in energy management optimization for microgrid. IEEE Innovative Smart Grid Technologies e Asia (Isgt Asia) (2013)
4. V.A. Graham, K.G.T. Hollands, A method to generate synthetic hourly solar radiation globally. Sol. Energy **44**(6), 333–341 (1990)
5. A.D. Duffie, W.A. Beckman, *Solar Engineering of Thermal Processes*, 4th edn. (Wiley, Hoboken, New Jersey, 2013).
6. M. Rohani, G. Nour, Techno-economical analysis of stand-alone hybrid renewable power system for Ras Musherib in United Arab Emirates. Energy **64**, 828–41 (2014)
7. M. Bhatia, N. Angelou, *Beyond Connections: Energy Access Redefined* (World Bank, Washington, DC, 2015).
8. System Adviser Model (SAM), NREL (2018). https://sam.nrel.gov/
9. Gas Turbine Magazine,https://gasturbineworld.com/. Accessed November 2019
10. Capstone Model C65 Performance; Technical Reference, Capstone Turbine Corporation, Chatsworth, CA, USA (2008)
11. Trading Economics (2019). https://tradingeconomics.com/ghana/interest-rate

Chapter 21
Recycling Mine Tailings for a Sustainable Future Built Environment

**Surya Maruthupandian, Napoleana Anna Chaliasou,
and Antonios Kanellopoulos**

Abstract The future sustainable built environment focuses mainly on environmental conservation and technological innovation and development. However, with infrastructure development, the consumption of raw materials such as cement, gypsum, sand, and stones increases. Therefore, use of industrial waste as raw material in construction shall be proposed as a sustainable and environment friendly alternative. Also, the higher demand for mineral commodities have led to increased mining and hence increased mining waste. The mine tailings being the wastes from rocks and minerals processing, are generally rich in Si, Ca, Al, Mg, and Fe, and also have considerable amounts of heavy metals and metalloids such as Pb, As, Co, Cu, Zn, V, and Cr. When tailings contain sulphide minerals, it may also lead to acid mine drainage. This makes the effective and efficient recycling and reuse of mine waste a major environmental concern. However, the physical, mineralogical and chemical composition of the mine tailings renders it a suitable material for use in civil engineering applications. This paper discusses the use of mine tailings of different origins for different civil engineering applications such as bricks, ceramics, fine aggregates, coarse aggregate and cementitious binders. This approach has a potential to reduce the demand on existing natural resources to face the demands of the exponentially developing infrastructure.

Keywords Sustainable construction · Mine wastes · Mine tailings · Reuse and recycle

21.1 Introduction

Environmental conservation and technological innovation and development are the major focus of future sustainable built environment [1]. Thus sustainable utilization of industrial wastes in civil engineering applications shall be a viable approach to

S. Maruthupandian (✉) · N. A. Chaliasou · A. Kanellopoulos
School of Engineering & Computer Science, Materials & Structures
Research Group, University of Hertfordshire, Hatfield, United Kingdom
e-mail: s.maruthupandian@herts.ac.uk

© The Author(s) 2021
I. Mporas et al. (eds.), *Energy and Sustainable Futures*, Springer Proceedings in Energy,
https://doi.org/10.1007/978-3-030-63916-7_21

meet the demands of raw materials such as cement, sand, aggregates, backfill soil and gypsum associated with rapid infrastructure development and spreading built environment.

Also, with the rapid development the demand for mineral commodities has increased leading to increase in mining operations. This leads to production of mining waste including waste rocks, over burden soil and mine tailings [2]. The fine-grained mineral waste left after removal of valuable material from ore is called mine tailing and is the major waste of mining processes [3]. This paper focuses on providing ways of utilization of mine tailings in different civil engineering applications.

21.2 Mine Tailings

The mining operations lead to a production of 20–25 billion tonnes of solid wastes every year, of which mine tailings constitute about 5–7 billion tonnes. [4]. These tailings are generally disposed, either by direct disposal of tailings in rivers, seas or disposal in form of slurry (25–30% solid) into a cell, confinement or dam [5]. Mine tailing management is a crucial issue as the physico-chemical characteristics of the wastes have severe impact on the soil and water, some of the more severe ones being acid mine drainage and leaching of heavy metals.

21.2.1 Properties of Mine Tailings

Mine tailings in general are crystalline, relatively loose and have a porous microstructure [6]. The specific gravity of the mine tailings was between 2.7–4.29 [7], [8] and the particle size of mine tailings varied amongst cement sized, silt sized and sand sized particles [9]. The water absorption was reported to be up to 7.15% [10] and pH values of the tailings were in near neutral or slightly alkaline region generally ranging between 6.69–10 [11–13]. The composition varied highly due to the variation in source mineral, process of extraction, mineral extracted and quality of the ore [14]. Most of the mine tailings were predominantly silicates and the distribution of major chemical oxides of different mine tailings is given in Fig. 21.1. The presence of heavy metals such as Pb, Zn, Cd, Cu, Ar, Cr, V, Zr poses a danger of leaching into the environment [15]. Mine tailings with sulphide minerals such as pyrite, arsenopyrite and pyrhottite [16] reacts with oxygen and water and cause acid mine drainage [9].

21.3 Research Trends for Application in Built Environment

Mine tailings are suitable for various civil engineering applications due to its physical and chemical properties. However, heavy metals leaching and acid mine drainage are

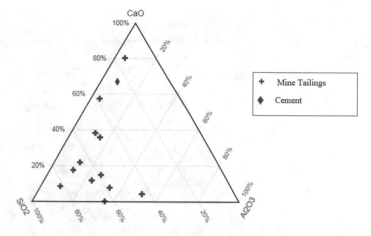

Fig. 21.1 Chemical Composition of Mine tailings

the major concerns associated with use of mine tailings. The past studies evaluated the feasibility and effect of use of mine tailings in various applications.

21.3.1   Bricks

Mine tailings can be used as an alternative to natural fine aggregates in brick production thus reducing the demand on sand mining in river beds. Fired clay bricks prepared using hematite mine tailings were found to be achieving a strength of 12.65 to 20.35 MPa and water absorption was found to be in range of 16.5–17.9% when the firing temperatures were varied from 850–1050 °C [14]. Optimum firing temperature was found to be 950–1000 °C [17]. Production of geopolymer bricks using copper mine tailing indicated that the strength increased with increase in molarity of the activator. However, the effective optimisation of water content and forming pressure plays a crucial role in pore structure and hence influences the compressive strength and water absorption of the bricks. [18]

21.3.2   Tiles

The minerals in mine tailings change from one phase to another with temperature. This property, makes mine tailing a suitable raw material for use in tiles and ceramics. When traditional tiles and ceramics impose increased demand on kaolinite and clay, use of mine tailings may prove to be an environment friendly and economical alternative. Kinnunen et al. [4] considered chemically bonded ceramics and geopolymerisation technology as two methods most suitable for effectively using mine tailings

Fig. 21.2 Furniture made of mine tailing in a heritage site in Portugal [20]

in ceramics. Hematite tailings proved to be an efficient fluxing agent in production of porcelain tiles, when used in percentages between 50 and 70%. Water absorption and porosity of the tiles decreased with increase in percentage of mine tailings and a strength up 72.5 MPa was achieved [19]. Gomez et al. [20] attempted the production of terrazzo tiles for decorative purposes using tungsten mine wastes and resin. The aged mine tailings were of a visually appealing ochre colour, whereas the heated tailing was reddish in colour owing to Iron oxidation. The samples were heated at 600 °C to obtain the desired colour. This colour variation makes application of mine tailings extendable use in heritage structures where the variation in colour of repair material is a major requirement. One such application in shown in Fig. 21.2.

21.3.3 Fine Aggregate

With the increased demand and shortage of quality fine aggregate, the alternative sources for fine aggregates are being explored globally [21]. Mine tailing as a replacement of fine aggregate in mortar and concrete has been studied widely, mainly due to its particle size and specific gravity. Borges et al. [22] reported that though the use of iron tailing as sand has no effect on strength properties, it affected the durability due to increased porosity and water absorption. Wang et al. [13] studied the freeze thaw performance of mortar samples with graphite mine tailings as fine aggregates. For short term cycles 40% replacement and for long term cycles 20% replacement performed better than control mixes. When gold mine tailings were used as a sand replacement a maximum strength of 47 MPa was achieved at 30% replacement. [10]

21.3.4 Coarse Aggregates

A very few studies have been carried out on use of mine tailings as coarse aggregate. Development of conductive concrete using graphite mine tailing as coarse aggregates was carried out by Liu et al. [23]. It was observed that strength reduced by about 29–42% for 15% replacement. The concrete with mine tailings performed well with respect to conductivity. The conductivity improved by about 50% for 15% replacement. This extends the applicability to de-icing/snow melting concrete for pavements in cold weather conditions and for applications where cathodic protection of reinforcement is required.

21.3.5 Cementitious Binders

Mine tailings may be used in cementitious binders as a raw material for clinker, as a supplementary cementitious material, as a direct cement replacement or as a precursor for alkali activated materials. For such applications, chemical and physical modification of mine tailings by thermal treatment, alkali activation and grinding may be performed to achieve desirable properties. Laura et al. [24] suggested that coal mine waste when activated, could be used as direct cement replacement. Mortar of 42.5 MPa and 32.5 MPa strength were obtained with 20 and 50% replacement and Ince [10] reported an optimum replacement percentage of 30%. Vargas and Lopez [25] reported that copper mine tailings when used as a supplementary cementitous material improved the mortar strength. Malagon et al. [26] used 7% of coal mining waste as a raw material for production of cement clinker and observed a 9–14% decrease in strength due to hindering of hydration by the Cu ions present in the mine waste. Calcination of tungsten mine waste with Na_2CO_3, followed by activation using sodium hydroxide solution yields high early age strength up to 45 MPa [27].

21.4 Discussions and Conclusions

The use of mine tailings have also been extended to other civil engineering applications such as cemented back fill [28], soil stabilization, landfill and embankments [29]. The past studies indicate the possibility of production of high strength cement and building components using mine tailings. The use of mine tailings in these building components and construction processes not only has the potential to reduce the demand on the existing natural resources but it could also provide an effective and environment friendly way of disposal of mine tailings. However, these applications have to be considered with caution taking into account the leaching of heavy metals and the possibility of acid mine drainage. With application of alkali activation [30], hydration [31], chemical bonding [4] such risks can be minimised and the mine tailings can be used with greater confidence.

References

1. Sustainable Built Environments | Vivian Loftness | Springer (2016)
2. G. Blight, Mine waste: a brief overview of origins, quantities, and methods of storage. *Waste*, 77–88 (2011)
3. D. Kossoff, W.E. Dubbin, M. Alfredsson, S.J. Edwards, M.G. Macklin, K.A. Hudson-Edwards, Mine tailings dams: Characteristics, failure, environmental impacts, and remediation. Appl. Geochemistry **51**, 229–245 (2014)
4. P. Kinnunen et al., Recycling mine tailings in chemically bonded ceramics—a review. J. Clean. Prod. **174**, 634–649 (2018)
5. J.S. Adiansyah, M. Rosano, S. Vink, G. Keir, A framework for a sustainable approach to mine tailings management: disposal strategies. J. Clean. Prod. **108**, 1050–1062 (2015)
6. L. Yu, Z. Zhang, X. Huang, B. Jiao, D. Li, Enhancement experiment on cementitious activity of copper-mine tailings in a geopolymer system. Fibers **5**(4), 1–15 (2017)
7. O. Onuaguluchi, Ö. Eren, Recycling of copper tailings as an additive in cement mortars. Constr. Build. Mater. **37**, 723–727 (2012)
8. F. Pacheco-Torgal, J.P. Castro-Gomes, S. Jalali, Investigations on mix design of tungsten mine waste geopolymeric binder. Constr. Build. Mater. **22**(9), 1939–1949 (2008)
9. R. Argane et al., Geochemical behavior and environmental risks related to the use of abandoned base-metal tailings as construction material in the upper-Moulouya district, Morocco. Environ. Sci. Pollut. Res. **23**(1), 598–611 (2016)
10. C. Ince, Reusing gold-mine tailings in cement mortars: mechanical properties and socio-economic developments for the Lefke-Xeros area of Cyprus. J. Clean. Prod. **238**, 117871 (2019)
11. J. Esmaeili, H. Aslani, Use of copper mine tailing in concrete: strength characteristics and durability performance. J. Mater. Cycles Waste Manag. **21**(3), 729–741 (2019)
12. A.M.T. Simonsen, S. Solismaa, H.K. Hansen, P.E. Jensen, Evaluation of mine tailings' potential as supplementary cementitious materials based on chemical, mineralogical and physical characteristics. Waste Manag. **102**, 710–721 (2020)
13. Z.R. Wang, B. Li, H.B. Liu, Y.X. Zhang, X. Qin, Degradation characteristics of graphite tailings cement mortar subjected to freeze-thaw cycles. Constr. Build. Mater. **234**, 117422 (2020)
14. Y. Chen, Y. Zhang, T. Chen, Y. Zhao, S. Bao, Preparation of eco-friendly construction bricks from hematite tailings. Constr. Build. Mater. **25**(4), 2107–2111 (2011)
15. Q. Wan, F. Rao, S. Song, R. Morales-Estrella, X. Xie, X. Tong, Chemical forms of lead immobilization in alkali-activated binders based on mine tailings. Cem. Concr. Compos. **92**(May), 198–204 (2018)
16. J. Kiventerä, L. Golek, J. Yliniemi, V. Ferreira, J. Deja, M. Illikainen, Utilization of sulphidic tailings from gold mine as a raw material in geopolymerization. Int. J. Miner. Process. **149**, 104–110 (2016)
17. C. Yang, C. Cui, J. Qin, X. Cui, Characteristics of the fired bricks with low-silicon iron tailings. Constr. Build. Mater. **70**, 36–42 (2014)
18. S. Ahmari, L. Zhang, Durability and leaching behavior of mine tailings-based geopolymer bricks. Constr. Build. Mater. **44**, 743–750 (2013)
19. Y. Chen, Y. Zhang, T. Chen, T. Liu, J. Huang, Preparation and characterization of red porcelain tiles with hematite tailings. Constr. Build. Mater. **38**, 1083–1088 (2013)
20. J.P. Castro-Gomes, A.P. Silva, R.P. Cano, J. Durán Suarez, A. Albuquerque, Potential for reuse of tungsten mining waste-rock in technical-artistic value added products. J. Clean. Prod. **25**, 34–41 (2012)
21. S.K. Kirthika, S.K. Singh, A. Chourasia, Alternative fine aggregates in production of sustainable concrete—a review. J. Clean. Prod., 122089 (2020)
22. P.H.R. Borges, F.C.R. Ramos, T.R. Caetano, T.H. Panzerra, H. Santos, Reuse of iron ore tailings in the production of geopolymer mortars. Rev. Esc. Minas **72**(4), 581–587 (2019)
23. H. Liu, K. Liu, Z. Lan, D. Zhang, Mechanical and electrical characteristics of graphite tailing concrete. Adv. Mater. Sci. Eng. **2018** (2018)

24. L. Caneda-Martínez, M. Frías, C. Medina, M.I.S. de Rojas, N. Rebolledo, J. Sánchez, Evaluation of chloride transport in blended cement mortars containing coal mining waste. Constr. Build. Mater. **190**, 200–210 (2018)
25. F. Vargas, M. Lopez, Development of a new supplementary cementitious material from the activation of copper tailings: Mechanical performance and analysis of factors. J. Clean. Prod. **182**, 427–436 (2018)
26. B. Malagón, G. Fernández, J.M. De Luis, R. Rodríguez, Feasibility study on the utilization of coal mining waste for Portland clinker production. Environ. Sci. Pollut. Res. **27**(1), 21–32 (2020)
27. F. Pacheco-Torgal, S. Jalali, Influence of sodium carbonate addition on the thermal reactivity of tungsten mine waste mud based binders. Constr. Build. Mater. **24**(1), 56–60 (2010)
28. J. Zheng, Y. Zhu, Z. Zhao, Utilization of limestone powder and water-reducing admixture in cemented paste backfill of coarse copper mine tailings. Constr. Build. Mater. **124**, 31–36 (2016)
29. I. Vegas, M. Cano, I. Arribas, M. Frías, O. Rodríguez, Physical-mechanical behavior of binary cements blended with thermally activated coal mining waste. Constr. Build. Mater. **99**, 169–174 (2015)
30. F. Rao, Q. Liu, Geopolymerization and its potential application in mine tailings consolidation: a review. Miner. Process. Extr. Metall. Rev. **36**(6), 399–409 (2015)
31. J. Nouairi et al., Study of Zn-Pb ore tailings and their potential in cement technology. J. African Earth Sci. **139**, 165–172 (2018)

Chapter 22
Optimal Design of Environmental-Friendly Hybrid Power Generation System

SeyedVahid Hosseini, Ali Izadi, Afsaneh Sadat Boloorchi, Seyed Hossein Madani, Yong Chen, and Mahmoud Chizari

Abstract Combination of both renewable and fuel-based generation systems is an advantageous approach to develop off-grid distributed power plants. This approach requires evaluation of the techno-economic potential of each source in a selected site as well as optimization of load sharing strategy between them. Development of a remote hybrid power plant in an off-grid area is the interest of this study. Defining all available combinations, characteristics of performance, cost and availability of them evaluated. Applying constraints, multi-objective target domain based on load following and Levelized Cost of Electricity is established in which by utilizing Pareto front approach, optimized scenarios is achieved.

Keywords Hybrid power generation · Solar · Wind · Energy · Micro gas turbine · Micro grid · Micro power plant

22.1 Introduction

For many under-development countries, hybrid generation systems are the solution which could satisfy various requirements of power generation including availability, renewability and cost. The authors presented a procedure to evaluate the performance and economy of a hybrid system of wind, photovoltaic (PV) and Micro Gas Turbine (MGT) with the required battery capacity in a remote area in South Africa as a test case in [1] and suggest optimization to size of a hybrid system of all resources.

The optimization of the power plant can be considered in two groups of conventional and new generation methods [2]. The conventional methods include the trade-off approaches, iterative approaches, linear and mixed integer linear programming.

S. Hosseini (✉) · Y. Chen · M. Chizari
Univesity of Hertfordshire, Hatfield, UK
e-mail: v.hosseini@herts.ac.uk

A. Izadi · S. H. Madani
Samad-Power Ltd, Milton Keynes, UK

A. S. Boloorchi
Turbotec Co, Tehran, Iran

© The Author(s) 2021
I. Mporas et al. (eds.), *Energy and Sustainable Futures*, Springer Proceedings in Energy,
https://doi.org/10.1007/978-3-030-63916-7_22

The new generation approaches, on the other hand, integrate mathematical models with the computer programs in the heuristic manner to obtain a solution [3].

To find the optimum size of a hybrid system with solar PV and wind generators, a harmony search method is employed in [4]. In this order, several configurations are defined as "Harmony". Sizes of solar PV generators and wind turbines, as well as estimated costs of the system, are specified in each harmony. Considering optimization targets, discrete search technique employed to find the solution.

In current study, Levelized Cost of Electricity (LCOE) [5] and load following of the generation system is considered as the target objects. The procedure is adopted in a rural area near Cape Town, South Africa and to apply direct optimization, solution domain of all scenarios using defined available sources is constructed. Various aspects of scenarios are investigated. Then, Pareto front technique is employed to obtain the optimum solution.

22.2   Theory and Methodologies

To define the solution domain, combinations of Capstone MGT30, MGT65 and MGT 200S with 0 to 4 wind turbine of 100 kW power are considered. The size of Solar PV in each configuration is obtained based on the assumption of providing at least 90% of the demand load in the worst month of each scenario. Therefore, 45 scenarios are extracted which are listed in Table 22.1.

Since there are a finite number of scenarios, the direct optimisation technique is employed to find the solution. In this method, techno-economic parameters of all scenarios are evaluated which make it possible to select the global optimal solution. Various parameters could be defined to show the performance, cost and availability of the system among which annual generation, Levelized Cost of Electricity (LCOE), load following deviation, nominal capacity, number of days with 1 hour shortage or blackout, and renewability of the hybrid generation system is considered to trim unpractical scenarios.

Table 22.1 Available scenarios

		Wind Turbine				
		0	100	200	300	400
MGT	0	2680	1520	610	180	0
	30	2200	1070	300	40	0
	65	1640	560	130	0	0
	30+65	1170	280	0	0	0
	2x65	610	80	0	0	0
	2x65+30	260	0	0	0	0
	3x65	40	0	0	0	0
	200	10	0	0	0	0
	200+30	0	0	0	0	0

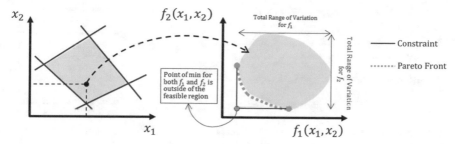

Fig. 22.1 The Pareto front in a multi-objective optimisation domain

LCOE are selected as the objectives for the optimization process that means the desired hybrid power plant not only have to track the demand load profile but also provide the electricity in minimum LCOE. To quantify the load following behaviour, least square of the difference between monthly load demand and generation is considered as the deviation parameter.

Based on the objectives' characteristics which are in contrary to each other the Pareto Front approach is applied to find the optimum. This method which is illustrated schematically in Fig. 22.1 generally employed when minimising one of the targets yields to increase of the other ones. Any of the scenarios which are places on the Pareto front could be selected as an optimum solution regarding the weight of each target for a project.

22.3 Results and Discussion

A rural area near Cape Town is considered as a case study in this article and extracted techno-economic parameters of all 45 scenarios are plotted in Fig. 22.2. It is obviously clear that some of the scenarios generate several times higher than the annual required load. Therefore, to achieve a better optimization domain, some of them are ignored by applying two following constraints:

- Size: Annual generation of the system should be less than 2,500,000 kWh.
- Availability: There should be no days with more than 1 h no generation situation.
- Renewability: Utilization of renewable sources should be more than 20% of overall generation.

Although the values for these constrain are specified, other values or constraints may be added due to the condition of a project. The remaining scenarios are indicated in Table 22.2. All of the scenarios with no MGT are eliminated because of the availability constrains and the others eliminated by size and renewability ones. In

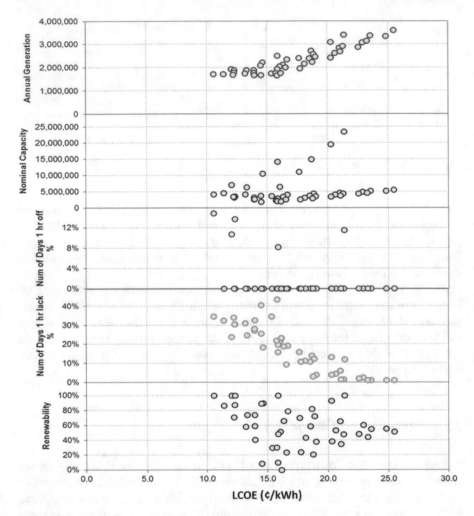

Fig. 22.2 Techno-economic parameters of all scenarios

remaining scenarios, some of them benefit from only two sources which means those do not include either any wind turbine or any solar PV. In Fig. 22.3 these two-source system are compared with the three-source systems in optimization targets domain which shows the effect of hybridization in improving both technical and economic aspects of the system.

Multi-objective optimization domain of the remaining 24 scenarios as well as the illustration of Pareto line is plotted in Fig. 22.4. Pareto front is passing through three following scenarios:

Table 22.2 Remaining scenarios after applying size, availability and renewability constraints

		Wind Turbine				
		0	100	200	300	400
MGT	0	2680	1520	610	189	0
	30	2200	1070	300	40	0
	65	1640	560	130	0	0
	30+65	1170	280	0	0	0
	2×65	610	80	0	0	0
	2×65+30	260	0	0	0	0
	3×65	40	0	0	0	0
	200	10	0	0	0	0
	200+30	0	0	0	0	0

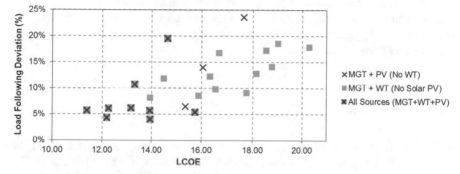

Fig. 22.3 Comparison between two-source and three-source systems

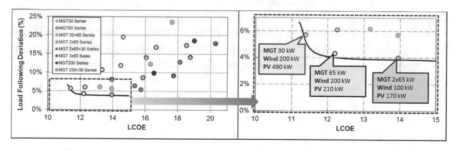

Fig. 22.4 Multi-objective domain and Pareto front

1- MGT 30 kW, WT 2 × 100 kW, PV 490 kW, Battery 1640 kWh
2- MGT 65 kW, WT 2 × 100 kW, PV 220 kW, Battery 1800 kWh
3- MGT 2 × 65 kW, WT 100 kW, PV 170 kW, Battery 1530 kWh

It could be seen that by increasing the size of the MGT, the load following behaviour of the system is improved but the LCOE is increased. Load following of these three optimum solutions is also plotted in Fig. 22.5 which shows more deviation in small MGT scenario.

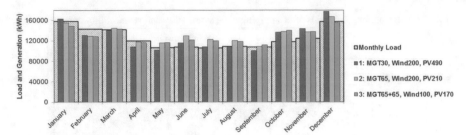

Fig. 22.5 Load following deviation of the selected scenarios

This Pareto front provides optimum solutions based on the selected targets as well as required data for decision makers. Having all optimum solution in hand, one may consider weights for each target and select the most appropriate one based on especial condition of the project.

22.4 Conclusion

Techno-economic parameters of all available solutions employing 30, 65 and 200 kW MGTs, as well as 0 to 4 sets of wind turbine in combination with appropriate solar PV, are evaluated. Applying size, availability and renewability constraints, some of the scenarios were eliminated. Among the remaining ones, load following behaviour and LCOE were considered as the target and the multi-objective Pareto front was obtained. It was shown that combining both three resources yields to enhancement of the overall techno-economic parameters of the system. The authors recommend that this procedure could be improved by employing a wider range of wind turbines. The criteria of suppling at least 90% of monthly loads to size the solar PV could also be investigated.

References

1. S.V. Hosseini, H. Madani, A. Izadi, M. Chizari, Design procedure of a hybrid renewable power generation system, in International Conference on Energy and Sustainable Futures (ICESF), Hatfield (2020)
2. W.L. Theo, J.S. Lim, W.S. Ho, H. Hashim, C.T. Lee, Review of distributed generation (DG) system planning and optimisation techniques: Comparison of numerical and mathematical modelling methods, Renewable and Sustainable Energy Reviews, vol. 67 (2017)
3. S. Twaha, M.A. Ramli, A review of optimization approaches for hybrid distributed energy generation systems: off-grid and grid-connected systems. Sustainable Cities and Soc. **41** (2018)

4. A. Askarzadeh, Developing a discrete harmony search algorithm for size optimization of wind–photovoltaic hybrid energy system. Solar Energy **98** (2013)
5. M. Bortolini, M. Gamberi, A. Graziani, F. Pilati, Economic and environmental bi-objective design of an off-grid photovoltaic–battery–diesel generator hybrid energy system. Energy Conver. Manag. **106** (2015)

Chapter 23
Distributed Activation Energy Model for Thermal Decomposition of Polypropylene Waste

S. Kartik, Hemant K. Balsora, Abhishek Sharma, Anand G. Chakinala, Abhishek Asthana, Mukesh Goel, and Jyeshtharaj B. Joshi

Abstract Thermal decomposition kinetics of Polypropylene (PP) waste is extremely important with respect to valorisation of waste plastics and production of utilizable components viz. chemicals, fuel oil & gas. The present research study focuses on pyrolysis kinetics of PP waste, which is present as a fraction of municipal plastic waste through distributed activation energy model (DAEM). The decomposition kinetics for PP follows a Gaussian distribution, where the normal distribution curves were centred corresponding to activation energy of 224 kJ/mol. The standard deviation of the distribution for the PP sample was found to be 22 kJ/mol indicating its wider distribution of decomposition range. The data validation has been carried out by comparing the rate parameter and extent of conversion values calculated through DAEM model with the Thermogravimetric analysis (TGA) experiments carried out for PP at various heating rates of 5, 10, 20 and 40 °C/min.

Keywords Pyrolysis · Plastics · TGA · DAEM · Gaussian distribution · Rate

S. Kartik · A. Sharma · A. G. Chakinala
Department of Chemical Engineering, Manipal University Jaipur, Dehmi Kalan, Jaipur 03007, Rajasthan, India

H. K. Balsora
Department of Chemical Engineering, Shroff S.R. Rotary Institute of Chemical Technology, Vataria, Ankleshwar, Bharuch 393135, Gujarat, India

A. Asthana · M. Goel (✉)
Department of Engineering and Mathematics, Sheffield Hallam University, Sheffield S1 1WB, UK
e-mail: mukesh.goel@shu.ac.uk

J. B. Joshi
Department of Chemical Engineering, Institute of Chemical Technology, Mumbai 400019, Maharashtra, India

© The Author(s) 2021
I. Mporas et al. (eds.), *Energy and Sustainable Futures*, Springer Proceedings in Energy,
https://doi.org/10.1007/978-3-030-63916-7_23

23.1   Introduction

Plastics are the most innovative materials of 20th century where their consumption is increasing everyday due to properties of mechanical strength, inertness and durability. The demand for plastic materials will be doubled by 2025 which will have an additional burden on raw materials used for their production, more specifically on fossil fuel resources [1, 2]. This increasing demand along with human lifestyle changes results in large proportion of mismanaged plastic waste which is likely to end up in sinks. Waste management plays a vital role in addressing accumulation of plastic waste considering its biodegradability, where option of pyrolysis is viable for plastics in terms of producing of liquid, gas and solid char fractions considering energy investments into the process.

Pyrolysis kinetics is important with respect to upscaling of polymer recycling process at commercial scale [3]. Here the relative rates of decomposition, cracking and other polymerisation reactions affect the quality and quantity of oil produced during the process. Thermogravimetry Analysis (TGA) is used to study decomposition kinetics and subsequently used for evaluation of kinetic parameters. The present research study focuses on pyrolysis kinetics of polypropylene (PP) through distributed activation energy model where the accuracy and versatility of DAEM assists for reproducing kinetic parameters considering the complex pyrolysis phenomena. The activation energies at various conversion levels is assumed to follow Gaussian distribution, in which the current study helps in the estimation of kinetic parameters. Subsequently estimated kinetic parameters can be directly used for the prediction of rate of decomposition of polypropylene at various heating rates thereby avoiding the need of performing pyrolysis experiments.

23.2   Experimental

Commercial grade PP samples were purchased from Reliance Industries Limited in which decomposition studies were carried out at heating rate of 5, 10, 20 and 40 °C/min. PP beads were crushed to a particle size of 0.5 mm which was subsequently subjected for further experimentations and TGA analysis in temperature range of 30–600 °C. The TGA experiments were carried out in a Perkin Elmer differential thermal analyser—Diamond TG/DTA model of Perkin Elmer, USA, under non-isothermal conditions at Sophisticated Analytical Instrument Facility in IIT Bombay.

The amount of sample used for analysis was 15 mg having a particle size of 0.5 mm. Nitrogen flow of 50 ml/min was maintained through sample during experiments. The reproducibility of results obtained from experiments was ensured by repeatedly analysing the sample three times.

23.2.1 Theoretical Considerations

The rate of decomposition encountered in solid state kinetics is dependant up on Temperature T, extent of conversion (α) and heating rate ($\beta = \frac{dT}{dt}$)

$$\frac{d\alpha}{dt} = \beta \frac{d\alpha}{dT} = k(T)f(\alpha) \tag{23.1}$$

The extent of conversion α, is given by

$$\alpha = \frac{m_0 - m_t}{m_0 - m_f} \tag{23.2}$$

where m_0, m_t and m_f specify initial mass, mass at a time 't' and final mass of the sample respectively.

The parameter $k(T)$ signifies rate as a function of temperature which is represented by Arrhenius equation.

$$k(T) = A exp\left(-\frac{E_a}{RT}\right) \tag{23.3}$$

Combination of Eq. (23.1) and (23.3) results in explicit expression of reaction rate which can be written as

$$\frac{d\alpha}{dt} = \beta \frac{d\alpha}{dT} = A exp\left(-\frac{E_a}{RT}\right)f(\alpha) \tag{23.4}$$

Rearranging Eq. (23.4) we will get

$$g(\alpha) = \int_0^\alpha \frac{d\alpha}{f(\alpha)} = \frac{A}{\beta} \int_0^T exp\left(-\frac{E_a}{RT}\right) dT \tag{23.5}$$

$$p(x) = \int_0^T exp\left(-\frac{E_a}{RT}\right) dT = \left(\frac{e^{-x}}{x}\right)\frac{(x^3 + 18x^2 + 86x + 96)}{(x^4 + 20x^3 + 120x^2 + 240x + 120)} \tag{23.6}$$

where $x = \left(\frac{E_a}{RT}\right)$

The temperature integral presented in Eq. 23.5 has no analytical solution, where fourth order Senum Yung approximation given in Eq. 23.6 is used to evaluate temperature integral. The rate of decomposition as represented by Eq. 23.4, where rate calculations are carried against temperature at similar extent of conversion at different heating rates. The representational form of Friedmann Iso-conversional method is given below.

23.2.1.1 Friedmann Method (FR)

Friedman's method [4] is a differential iso-conversional method is obtained by taking logarithm on both sides of Eq. (23.4)

$$\ln\left(\beta\frac{d\alpha}{dT}\right) = \ln(k_0 f(\alpha)) - \left(\frac{E_a}{RT}\right) \tag{23.7}$$

Plotting $\ln\left(\beta\frac{d\alpha}{dT}\right)$ or $\ln\left(\frac{d\alpha}{dt}\right)$ against $\left(\frac{1}{T}\right)$ over the entire range of conversion will yield the value of activation energy.

23.2.1.2 Distributed Activation Energy Model

The distributed activation energy model assumes that decomposition of material is carried out through large number of independent reactions, each of them is having its own activation energy and frequency factor. It is further assumed that reactivity distribution follows a Gaussian distribution representing continuous distribution of activation energy. The rate of reaction represented in terms of this continuous distribution F(E) as

$$\frac{d\alpha}{dT} = \int_0^\infty \frac{k_0}{\beta} \exp\left[-\frac{E}{RT} - \frac{k_0}{\beta}\int_0^T \exp\left(-\frac{E}{RT}\right)dT\right]F(E)dE \tag{23.8}$$

The true distribution of activation energy F(E) is represented through Gaussian distribution with mean activation energy E_0 and having standard deviation of σ is mentioned below.

$$F(E) = \frac{1}{\sqrt{2\pi}\sigma}\exp\left[-\frac{(E - E_0)^2}{2\sigma^2}\right] \tag{23.9}$$

The comparison of experimental and simulated data is represented through minimization of objective function. (O.F)

$$O.F = \sum_{i=1}^{n_d}\left[\left(\frac{d\alpha}{dT}\right)_{i,exp} - \left(\frac{d\alpha}{dT}\right)_{i,cal}\right]^2 \tag{23.10}$$

23.3 Results and Discussion

The TGA curves for polypropylene are shown in Fig. 23.1. Thermograms are indicative of single stage devolatalization in temperature range of 300–600 °C, where it is characteristic of higher volatile content in the PP sample. The peaks in dTG curve

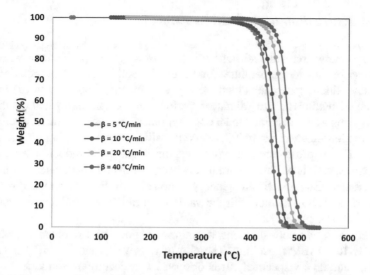

Fig. 23.1 Weight loss and dTG curves for PP at different heating rates of 5, 10, 20 and 40 °C/min

shown in Fig. 23.2 indicate that reaction rate for solid state decomposition reaches a maximum at some intermediate stage of conversion. The temperatures for this maximum at different heating rates of 5, 10, 20 and 40 °C/min found from dTG peaks are 442, 453, 465, 481 °C for PP. This indicates the temperature level at which reactor has to be operated in order to maximise the conversion of feed plastics in a continuous reactor system.

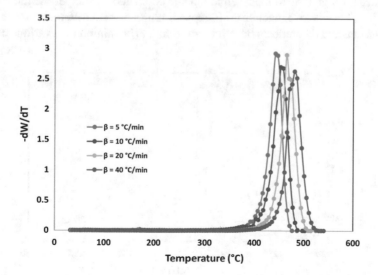

Fig. 23.2 dTG curves for PP at different heating rates of 5, 10, 20 and 40 °C/min

23.3.1 Estimation of Activation Energy

Activation energy remains constant for most of single step reactions where rate
constant is related to temperature. However, for solid state decomposition reactions, these kinetic parameters tend to vary with extent of conversion (α) [5]. Iso-
conversional methods help to estimate the value of activation energy without the prior
knowledge regarding reaction model or to hypothesize a form of kinetic equation, as
any conversion function can fit TGA decomposition data by varying kinetic parameters [6]. This is presented as a significant disadvantage of model-based methods
which rely on results of kinetic parameters obtained from Iso-conversional methods
[7]. The mean value of activation energy obtained from Friedmann isoconversional
plot for PP is 224.34 kJ/mol with the value of correlation coefficient representing
exactness of fit are 0.9931.

The E_α vs α and corresponding k_0 values obtained by Friedman's method for
PP is used for DAEM modelling. The plot of E vs V/V* and E vs f(E) is given in
Figs. 23.3 and 23.4 respectively. It is observed that decomposition kinetics for PP
follows a Gaussian distribution, where the normal distribution curves were centred
corresponding to activation energy of 224 kJ/mol corresponding to the level for extent
of conversion of 0.6. The standard deviation (σ) for PP waste sample was found to
be 22 kJ/mol which is indicative of its wider decomposition range.

Larger values of σ represent broader reaction profiles at constant heating rate and
these parameters vary with respect polymers due to their structural heterogeneity. The
isoconversional Friedmann method is used to estimate the value of activation energy
and frequency factor at different levels of conversion. With the Gaussian distribution
assumed for variation of activation energy represented by Eq. 23.9 the value of rate is
calculated through DAEM model equation and compared with experimental value of
rate $\left(\frac{d\alpha}{dT}\right)$. The objective function (O.F) representing square of difference between the
experimental and theoretical rate values are found to be minimum at various heating

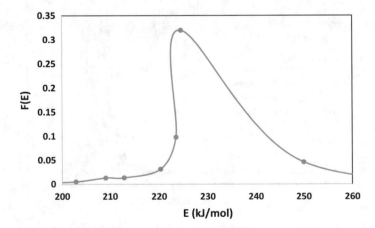

Fig. 23.3 F(E) curve estimated for PP

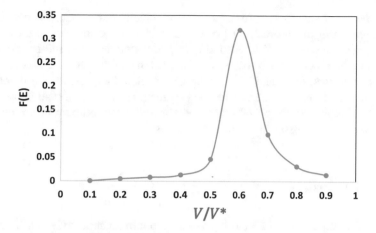

Fig. 23.4 F(E) vs V/V* relationship for PP

rates of 5, 10, 20 and 40 °C/min, which indicates that prediction of rate through DAEM model equation is closer to that of experimental values. The comparison between theoretical and experimental plots are given in Fig. 23.5 for various heating rates investigated in this manuscript.

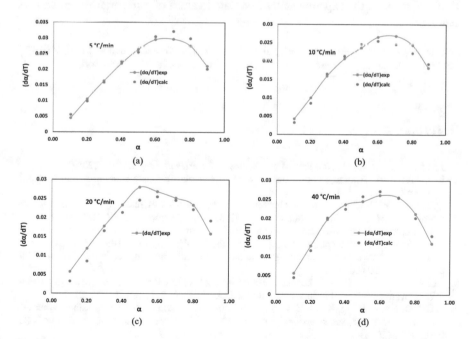

Fig. 23.5 Comparison of rate from experimental and prediction of DAEM for PP at various heating rates of 5 °C/min, 10 °C/min, 20 °C/min and 40 °C/min

The practical significance for estimation of kinetic parameters through DAEM involve predicting the rate of decomposition considering the complex nature of pyrolysis degradation process. Estimated kinetic parameters can be used for prediction of degradation model, which will be very helpful in dealing with the plastic wastes. In addition to this, the identified decomposition model expressed as a function of temperature and heating rate can be used to identify the operational temperature, where continuous reactor operation can be ensured with optimum heat input. This will prevent energy losses in tackling the wastes.

23.4 Conclusion

Thermal decomposition of PP has been carried out in present study at heating rates of 5, 10, 20 and 40 °C/min. The decomposition data so obtained is processed through Friedmann Isoconversional method to calculate the value of Activation Energy and Frequency factor at various conversion levels. Distributed Activation Energy Modelling (DAEM) was applied for decomposition of PP to calculate the rate of degradation, assuming Gaussian distribution for Activation Energy at various conversion levels. The normal distribution curves were centred corresponding to activation energy of 224 kJ/mol for the level of conversion of 0.6. The DAEM model is validated through the experimentally calculated values of rate parameter at various heating rates.

Acknowledgements Authors are thankful to Bharuch Enviro Infrastructure Limited (BEIL) Ankleshwar, India for providing the financial support to carry out this research work.

References

1. J.M.H. Army Lusher, P. Hollman, Microplastics in fisheries and aquaculture (2017)
2. L. Lebreton, A. Andrady, Future scenarios of global plastic waste generation and disposal, Palgrave Commun., 1–11 (2019)
3. Q.V. Bach, W.H. Chen, Pyrolysis characteristics and kinetics of microalgae viathermogravimetric analysis (TGA): A state of art review. Bioresource Technol. **246**, 88–100 (2017)
4. H.L. Friedman, Kinetics of thermal degradation of char-forming plastics from thermogravimetry. application to a phenolic plastic. J. Polym. Sci. Part C Polym. Symp **6**, 183–195 (1964)
5. P. Simon, Isoconversional methods—fundamentals meaning and application. J. Therm. Anal. Cal. **76**, 123–132 (2004)
6. J.E. White, W.J. Catallo, B.L. Legendre, Biomass pyrolysis kinetics: a compartive critical review with relevant agricultural residue case studies. J. Appl. Anal. Pyrol. **91**, 1–33 (2011)
7. J. Cai, D. Xu, Z. Dong, X. Yu, Y. Yang, S.W. Banks, A.V. Bridgwater, Proceessing of thermogravimetric data for Isoconversional kinetic analysis of lignocellulosic biomass pyrolysis **82**, 2705–2715 (2018)

Chapter 24
Innovative Strategy for Addressing the Challenges of Monitoring Off-Shore Wind Turbines for Condition-Based Maintenance

Amin Al-Habaibeh, Ampea Boateng, and Hyunjoo Lee

Abstract Off-shore wind energy technology is considered to be one of the most important renewable energy source in the 21st century towards reducing carbon emission and providing the electricity needed to power our cities. However, due to being installed away from the shore, ensuring availability and performing maintenance procedures could be an expensive and time consuming task. Condition Based Maintenance (CBM) could play an important role in enhancing the payback period on investment and avoiding unexpected failures that could reduce the available capacity and increase maintenance costs. Due to being at distance from the shore, it is difficult to transfer high frequency data in real time and because of this data transferring issue, only low frequency-average SCADA data (Supervisory Control And Data Acquisition) is available for condition monitoring. Another problem when monitoring wind energy is the massive variation in weather conditions (e.g. wind speed and direction), which could produce a wide range of operational alerts and warnings. This paper presents a novel case study of integrated event-based wind turbine alerts with time-based sensory data from the SCADA system to perform a condition monitoring strategy to categorise health conditions. The initial results presented in this paper, using vibration levels of the drive train, indicate that the suggested monitoring strategy could be implemented to develop an effective condition monitoring system.

Keywords Wind turbines · Condition monitoring, Condition-Based maintenance · CBM · SCADA

A. Al-Habaibeh (✉)
Product Innovation Centre, Nottingham Trent University, Nottingham, UK
e-mail: Amin.Al-Habaibeh@ntu.ac.uk

A. Boateng · H. Lee
Offshore Renewable Energy Catapult, Blyth, Northumberland, UK

© The Author(s) 2021
I. Mporas et al. (eds.), *Energy and Sustainable Futures*, Springer Proceedings in Energy,
https://doi.org/10.1007/978-3-030-63916-7_24

24.1   Introduction

Due to the growth in the world's population, worldwide energy demand is on the increase [1]. The Paris Agreement to lessen the impact of climate change sets the target of keeping the global temperature increase below 2 °C of the pre-industrial stage [2]. The UK Government's Climate Change Act (2008) creates a target of reducing greenhouse gas emissions to 80% of the 1990 level by the year 2050 [3], this focuses the attention on renewables as the way forward. Wind turbine technology is well established for onshore and offshore. Onshore wind energy systems have their own limitations, particularly near homes, in relation to noise, shadow flicker; and also, in the countryside due to the possible effect on wildlife and the potential negative aesthetic impact [4]. Hence, this makes offshore wind energy farms a suitable option due to scalability and the reduced effect on residential areas and wildlife. But this comes with its own challenges, as the cost of maintenance could be expensive and the reduction in capacity due to faults could be significant. According to [5], 28% of installed wind turbine capacity in Europe is expected to be older than 15 years by 2020. As wind turbines reach the end of their designed service life, maintenance of aging assets become critical to ensure return on investment and availability of capacity. This makes CBM an important concept to ensure maintenance can be done with suitable TPM (Total Productive Maintenance) strategy. A recent survey of CBM of wind turbines have indicated that new models and methodologies are needed which will allow adaptive maintenance scheduling and the prediction of time-to-failure through prognostics and health asset management [6].

24.2   The Case Study

Figure 24.1 presents a schematic diagram of the offshore wind energy turbine used in this case study. The wind turbine in consideration is a 7 MW system with rotor diameter of circa 171 metres and standard hub height of about 110 m above mean sea level. It has rotation direction of clockwise looking downwind with minimum rotor speed of 5.9 rpm and rated rotor speed of about 10.6 rpm. The cut-in wind speed is 3.5 m/s and the cut-out wind speed is 25 m/s with rated wind speed of 13 m/s. The three blades are bolted on a pitch bearings which are connected to a hub. The hub is attached to the main low speed shaft to transfer the torque to the gear box. The gearbox increases the speed of the output shaft to a suitable rpm to power the generator to produce three phase electricity. Blade pitch angles are continuously controlled depending on wind speed conditions above the rated speed. It also provides an aerodynamic braking when stopping procedures of the control system is activated. The main shaft and the gearbox are rigidly supported by two main bearings. The main shaft takes the torque and other loads from the rotor and its supporting configurations and transmits the pure torque to the gearbox. The system has an integrated and compact drive train design with oil lubrication for the gearbox and the main bearings

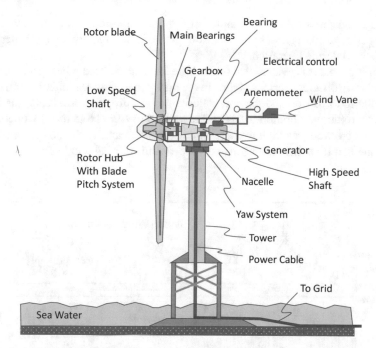

Fig. 24.1 A schematic diagram of the offshore wind energy turbine used in this case study

to enhance reliability. Rotor brake unit will stop the rotor speed following a decelera-tion by the aerodynamic braking systems via the orientation of the blades. Also there is a rotor locking system using with automatic or manual actuation. The generator, with nominal rotation speed of 400 rpm, is based on permanent magnet generator to provide higher efficiency. The electric generator is cooled using air-to-water cooling system and connected to the grid via a convertor.

24.3   Condition Monitoring and SCADA Data

The time history data provided by the SCADA system includes variables such as time, rotor speed, forces and torques at the hub centre and vibration of the gearbox. For ambient conditions, variables such as air temperature, wind direction and speed are captured, including nacelle temperature. For the grid, voltage, current, power and frequency are monitored. For the gearbox, sensors to monitor vibration, oil level, oil temperature and oil pressure are installed. Sensors for stress levels and vibration are also installed on bearings and near blade roots. The average data from the sensors are captured in the form of sensory feature characteristic (SCF) such as average, maximum, minimum and standard deviation. SCADA fault or waning system will also register the error of fault at the moment it happens via the control as ON or OFF status.

Figure 24.2 presents some examples of the data captured during operation (as a 10 min period) and Fig. 24.3 presents examples of how sensory data and alert data are registered during operation.

From Fig. 24.2 it is evident that the turbine has experienced wind speed less than 3.5 m/s during different periods of time in the figure (e.g. sample number 3500) which caused the rotor to stop and the power generation to drop to zero. It is clear that if there is no rotation, vibration levels will also drop to zero. Note that the fluctuation in wind speed produces high average vibration levels and in the power generated. But during that time there has been many alerts and warning within the independent SCADA system.

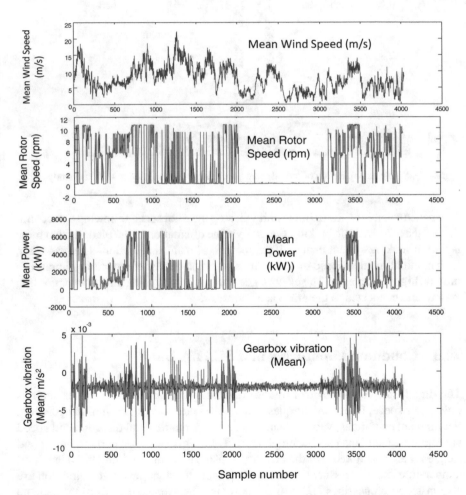

Fig. 24.2 Examples of the data in relation to wind speed, rotor speed, power generated and gearbox vibration

	Time Related Sensory Signals				
x_StartTime	WindSpeed_mps_Mean	WindSpeed_mps_Stdev	RotorSpeed_rpm_Mean	RotorSpeed_rpm_Stdev	NacelleOrientation_Deg_Min
07/01/2019 00:00	6.370099441	0.753717761	8.966484134	0.227632815	216.112
07/01/2019 00:10	6.180607961	0.714575458	7.198205017	1.658076885	224.0059
07/01/2019 00:20	6.331067486	0.538305239	5.596074492	0.129179038	224.0072
07/01/2019 00:30	6.72016821	0.573542969	5.593674458	0.131812148	224.0072
07/01/2019 00:40	6.942188706	0.77170168	7.71696081	1.634554206	224.0059
07/01/2019 00:50	6.771846741	0.894997826	8.994795418	0.222648755	224.0059
07/01/2019 01:00	7.484857397	0.819880329	8.984588782	0.210233846	224.0059
07/01/2019 01:10	7.054135905	0.893530656	8.97651875	0.209021978	224.0059
07/01/2019 01:20	7.004002306	1.091950075	8.96772136	0.207649735	224.0059
07/01/2019 01:30	7.68446681	1.115288492	9.018011423	0.234656298	224.0059
07/01/2019 01:40	7.630248966	1.012675852	8.982554951	0.188780785	224.0059
07/01/2019 01:50	7.733771401	1.060284336	9.040768199	0.21788103	224.0046
07/01/2019 02:00	8.988599418	1.081412888	9.898310209	0.415122059	224.0059
07/01/2019 02:10	8.482802151	1.063256988	9.412734406	0.321010496	224.0059

	Event Related Fault/Alert Signal				
ID	EVENT_TI_ME_GMT	SCADA_SENSO_R_TIME	COD_E	VALUE	DESCRIPTION
					(DEMOTED) Gearbox Filter Manifold Pressure 1
4046082	08:13.6	07/01/2019 00:00	T568	Off	Shutdown
4046098	56:07.7	07/01/2019 03:50	T624	On	IPR Relay Pick Up
4046099	56:09.7	07/01/2019 03:50	T624	Off	IPR Relay Pick Up
4046100	48:28.8	07/01/2019 04:40	T317	On	Inst High Side-Side Vib
4046101	53:42.6	07/01/2019 04:50	T1617	On	Noise shutdown
4046102	54:41.7	07/01/2019 04:50	T1617	Off	Noise shutdown
4046103	59:24.7	07/01/2019 04:50	T1617	On	Noise shutdown
4046104	59:56.6	07/01/2019 04:50	T1617	Off	Noise shutdown
4046105	01:01.9	07/01/2019 05:00	T1617	On	Noise shutdown
4046106	30:04.3	07/01/2019 05:30	T1617	Off	Noise shutdown
4046108	13:30.1	07/01/2019 07:10	T356	On	PCH Band 6 vibration warning
4046109	13:32.1	07/01/2019 07:10	T356	Off	PCH Band 6 vibration warning
4046110	13:47.5	07/01/2019 07:10	T356	On	PCH Band 6 vibration warning
4046111	13:50.5	07/01/2019 07:10	T356	Off	PCH Band 6 vibration warning

Fig. 24.3 Example of the time related sensory signals and Event related fault/alert signal

24.4 Sensory and Alarm Integration Procedure

In order to integrate the sensory data at a specific interval of every 10 min for example (see Fig. 24.4a), and the alert or error messages of the SCADA system (Fig. 24.4b), which could happen and be deactivated at any time, this paper suggests an integration process outlined by a generic example and articulated in Fig. 24.4c, d.

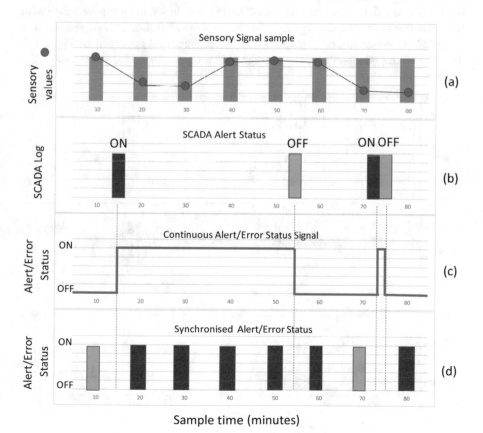

Fig. 24.4 Synchronisation principle of SCADA Alert/Error Status and Sensory signals

Based on the sensory data of Fig. 24.4a and the event-based warning in Fig. 24.4b, c presents a continuous function that is generated from the discrete alert system of Fig. 24.4b. The principle of this approach is that the alert or fault status is assumed ON or active, until an OFF or deactivation signal if found. If during the same time period the alert or error is activated and then deactivated, in this case the alert status is assumed active for that period and then deactivated for the following periods. Hence, the principle of the integration is: if the last status of the alert is active, then the alert will continue for the following periods until deactivation is observed. If activation and deactivation happen in the same period, then for that period the status will be active and will be considered re-activated for the following period, see Fig. 24.4.

24.5   Discussion: Time-Based and Event-Based Data Integration Case Study

Figure 24.5 presents an experimental case study to test the algorithm of the data integration to transfer a vibration alert which is event-based to be synchronised with time-based associated data. Figure 24.5a presents the alert time as it was transferred from SCADA system and processed using the suggested approach of Fig. 24.4. To validate the methodology, the average vibration sensor of the gearbox and the

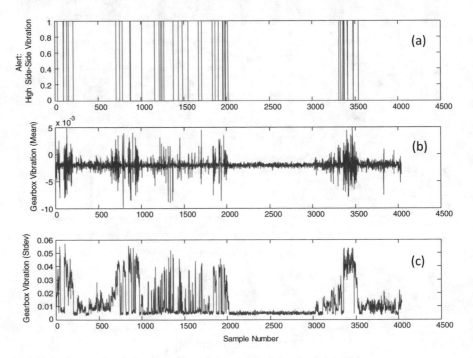

Fig. 24.5 Case study of vibration alert and the vibration signals of the gearbox

standard deviation of the vibration levels, Fig. 24.5b, c respectively, are compared with the transferred and processed alert/error signal. It is evident that the integration was synchronised as shown in Fig. 24.5 and the alert/error occurs when the vibration level of the gearbox reaches higher than norm levels. When comparing Fig. 24.2 with Fig. 24.5, it is evident that fluctuation in wind speed and rotor speed is what is causing the alert and the high level of vibrations.

24.6   Conclusion

This paper has suggested an integration methodology which allows the development of a Condition-Based Monitoring system by transferring event-based system errors and alerts to the time domain to be synchronised with the sampled data from the sensors. From the presented data, high fluctuation in wind speed is found to cause fluctuation in power generation and rotor speed which would cause high vibration levels in the drive train and hence a vibration alert. Future work will involve using Artificial Intelligence techniques to learn from experience to allow self-learning diagnostic of health conditions and future prognostics of wind turbines.

Acknowledgements The authors would like to thank Innovate UK and EPSRC for funding this research work; grant Reference EP/S515711/1.

References

1. U.S. Energy Information Administration, "International Energy Outlook 2017," 2017. Accessed: Nov. 28, 2017. [Online]. https://www.eia.gov/outlooks/ieo/pdf/0484(2017).pdf
2. United Nations Framework Convention on Climate Change, "ADOPTION OF THE PARIS AGREEMENT—Paris Agreement text English." United Nations, Paris, pp. 1–27, 2015, Accessed: Jul. 19, 2017. [Online]. https://unfccc.int/files/essential_background/convention/application/pdf/english_paris_agreement.pdf
3. *Climate Change Act*. London: The Parliament of the United Kingdom, pp. 1–103 (2008)
4. Heather Thomson, Willett Kempton, Perceptions and attitudes of residents living near a wind turbine compared with those living near a coal power plant, Renewable Energy, Volume 123, pp. 301–311, ISSN 0960-1481 (2018)
5. Suzan Alaswad, Yisha Xiang, A review on condition-based maintenance optimization models for stochastically deteriorating system, Reliability Engineering & System Safety, Volume 157, pp. 54–63 (2017), ISSN 0951-8320. https://doi.org/10.1016/j.ress.2016.08.009

Part IV
Electric Vehicles and Transportation Technology

Chapter 25
Impact of Replacing Conventional Cars with Electric Vehicles on UK Electricity Grid and Carbon Emissions

George Milev, Amin Al-Habaibeh, and Daniel Shin

Abstract This paper estimated the effect of electric vehicle transition on UK road and how it impacts on electricity supply and the reduction of carbon emissions. It used a scenario in which all cars that utilise internal combustion engines will be replaced by EVs in the UK. The methodology is based on speculating the future number of EVs in Great Britain, which helped in estimating the amount of additional electricity usage that would be required for the scenario. The results revealed that approximately 81 TWh of additional electricity must be produced annually to compensate for such expansion of EV. With that increase in electricity generation, the levels of carbon emissions from the electrical grid will rise slightly, by about 8.6 million tonnes of carbon dioxide per year. Given that combustion vehicles contribute to about 42% of the carbon emissions from the transport sector in the UK, it is concluded that the total amount of CO_2 in the country will decrease by approximately 12% of all cars with internal combustion engines are replaced by electric vehicles.

Keywords Electric · Car · Environment · Carbon · Emission

25.1 Introduction

Over the past few decades, the technology around transportation has evolved and shown dramatic developments. With the UK government pushing to go carbon neutral by 2050 [1], the demand for Electric Vehicle (EV) has had a huge rise more recently. This may due to the tendency of government legislation beginning to restrict the use of combustion vehicles (CV) on the road in order to further reduce Carbon Emission (CO_2). However, electricity is an energy source for EV which also produces CO_2 but much less than burning combustion engines. Hence, it is crucial to estimate the CO_2 reduction resulting from increasing the transition to EV. This study, therefore, analysed the reduction of CO_2 levels by comparing the current electricity usage in the UK against the scenario of UK road making 100% EV transition.

G. Milev (✉) · A. Al-Habaibeh · D. Shin
Department of Product Design, Nottingham Trent University, Nottingham, UK
e-mail: george.milev@ntu.ac.uk

© The Author(s) 2021
I. Mporas et al. (eds.), *Energy and Sustainable Futures*, Springer Proceedings in Energy,
https://doi.org/10.1007/978-3-030-63916-7_25

25.1.1 Description of CV and EV

The first modern petrol internal combustion engine (ICE) was constructed in the 1880s [2]. The difference between electric and ICE vehicle is that conventional cars burn fuel in the ICE, whereas EV store energy into a battery, which powers an electric motor [3]. However, assembling the combustion vehicle (CV) was a complex process. Therefore, at the end of the 19th century, EV was more favoured by people, for example, 90% of taxicabs were electric in 1899's New York [4].

In CV, the mixture of fuel and air is ignited by the sparks, causing a small explosion, resulting in downward movements of the engine's pistons. One of the by-products of the burnt fuel is carbon dioxide, which is released in the atmosphere during this process. It has been estimated that on average, a CV emits about 122 carbon dioxide per kilometre [5]. For EV, the electric energy is stored in the battery of the car and it provides power to the car's controller. The controller transfers that energy to the electric motor, which generates horsepower to spin the wheel [3]. Some models have a motor that is installed on each axle [6]. In research studying the performance of both CV and EV, it was found that EVs are 3.6 times more efficient in energy consumption [7]. EV use 3.4 times less total energy while driven compared to ICE vehicles, and also that combustion-powered cars emit 4.5 times more carbon emissions while in use [7].

25.1.2 EV Market Size

According to the latest reports of the UK's Department for Transport, there were a total of 38.9 million vehicles in the country, of which 82.43% (32 million) were cars and taxis [8]. Currently, in the UK there are more than 298,000 electric car models on the roads. Recent figures show that plug-in battery vehicles make up about 34% of the total new vehicle registrations in Great Britain [9]. When it comes to emissions popularity and energy/fuel consumption, the popularity of vehicles plays an important role.

Table 25.1 summarises the popularity of EV in the UK and their energy consumption. Tesla Model 3 and Nissan Leaf are the most popular electric vehicles in the country. Annually, all cars are traveling about 415 billion kilometres, or approximately 12,968.75 km on average [8].

25.1.3 Electricity Demand and Carbon Emissions in the UK

For the last 6 years, the electricity consumption rate in the UK has been dropping, particularly for households due to milder winters in recent years [10]. According to UK government data, households consume approximately 105 TWh of electricity

Table 25.1 Most popular EV in the UK [8]

Brand and model	Registered units in 2019	Popularity (%)	Energy consumption (kWh/100km)
Tesla Model 3	5,500	20.37	25
Nissan Leaf	4,500	16.67	16
Jaguar I-Pace	3,500	12.96	22.7
BMW i3	3,400	12.59	18.6
Volkswagen e-Golf	2,900	10.74	16
Renault Zoe	2,900	10.74	16
Nissan E-NV200	1,500	5.56	20.5
Tesla Model S	1,400	5.19	16.3
Tesla Model X	1,400	5.19	24

Fig. 25.1 Electricity
generation mix in the UK
[12]

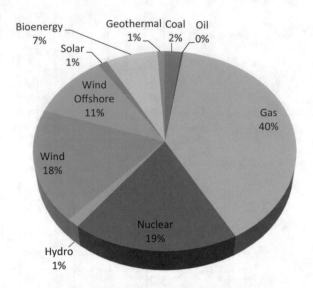

per year, which is reduced from 115 TWh since 2012 [10]. In energy supply, it was measured that the annual electricity production was 307 TWh in 2019, which is reduced from 325.17 TWh in 2012 [11]. In recent years, the UK's energy generation mix has reduced the use of coal and increased the usage of more renewable resources (Fig. 25.1).

Coal and gas contribute to a high amount of CO_2 in the UK. However, CO_2 level has been reducing for the last 5 years due to the reduction of using these energy sources and by increasing the use of renewable sources and nuclear power. The transport sector in Great Britain contributes to about 33% of the carbon dioxide emissions (119.6 $MtCO_2$) in 2019, 11% of reduction compared to 134.3 $MtCO_2$ in 2005 [14]. Power stations emit approximately 15% (57.4 $MtCO_2$/year) of the total carbon emissions [14]. It has been estimated that EV produces about 48 g of carbon emissions during the manufacturing stage, compared to 31 g for CV, but in the longer-term, CV produce significantly more emissions from burning fuels [7].

25.2 Methodology

Through the literature review, the current study acquired the required parameters to estimate the amount of electricity needed in supply and how much of CO_2 reduction can be generated for the scenario of electrifying all cars on UK road. In order to retrieve the proposed estimation, following 4 equations were set:

The study began by determining the **average energy consumption (AEC)** of EVs according to their brand and model popularity via using Eq. (25.1).

$$\sum_{i=1}^{n} = (EC_i \times PP_i) = \text{AEC} \tag{25.1}$$

EC represents the **electricity consumption** of each vehicle brand and model (kWh/100 km) and **PP** is the **percentage proportion** of electric cars according to their popularity, it is represented as (value)%/100; **n** is the **number** of electric car brands and models included in our study that is 9 (Table 25.1).

The next stage was to calculate the total electricity needed based on the EV transition scenario, using Eq. 25.2:

$$ER = (ATD \div 100) \times AEC \tag{25.2}$$

ER represents the **energy required** based on the EV transition scenario in kWh, later converted to TWh; **ATD** is the **annual traveling distance** for all cars in the UK (km); **AEC** is the **average energy consumption** of EV (kWh/100 km). The calculation also considered that there are about 7% (**ER + 7%**) distribution losses through the electricity grid in the UK [15] (Table 25.2).

In measuring the carbon emission, the current study estimated the emissions from cars using the following Eq. 25.3:

$$Em = ATD \times CI \tag{25.3}$$

where; **Em** is the **carbon emissions** released by all cars in the UK (kg), later converted to million tonnes; **ATD** is the **average traveling distance** (km); **CI** is the **carbon intensity** of the cars (kgCO$_2$/km).

Since EV does not produce any emissions while driven, the value for **Em**, which represents the amount of Carbon dioxide that all cars in the UK emit annually, was subtracted from current CO2 emission on the road today. Both current and newly estimated road CO$_2$ levels were compared. The final step was to estimate the changes

Table 25.2 Carbon intensity of electricity's sources [13]

Source	gCO$_2$/kWh
Coal CCS	220
Oil	314
Gas CCS	200
Nuclear	26
Hydro	7
Wind	15
Solar PV	88
Bioenergy	165
Geothermal	15

of CO_2 under the EV transition scenario. The **ER** value from Eq. 25.2 was distributed accordingly to the UK's energy generation mix (i.e. 40% for gas, 19% for nuclear, etc.) This is due to each source of electricity emits a different amount of carbon emissions per kWh of electricity produced. Equation 25.4 was used to calculate the increased CO_2 from each source and then they were summed to find the total amount.:

$$CO_2 = ES \times CI \tag{25.4}$$

CO_2 is **carbon dioxide emissions** ($kgCO_2$), later converted to $MtCO_2$; **ES** is the **electricity generated by each source** (kWh); **CI** is the **carbon intensity** of each corresponding source ($kgCO_2/kWh$). The total difference in annual carbon emissions based on the scenario was measured.

25.3 Results and Discussion

Using Eq. 25.1 it was calculated that the average energy consumption (AEC) by EV according to their brand and model popularity is 19.45 kWh/100 km.

Using the Eq. 25.2, it was estimated that the energy required for the EV scenario (ER) per annum would be 86.3 TWh which added the 7% grid losses, leading to a total of 22% rise in energy production at the supply phase.

In relation to carbon emissions (Eq. 25.3), the result revealed 49.5 $MtCO_2$/year (Em) are emitted by all CV today. EV does not produce any traffic emissions, which will lead to a 41.4% reduction in transport CO_2 levels per year.

Using Eq. 25.4, we estimated that a total of 8.6 $MtCO_2$/year will need to be generated additionally from the electricity grid for a 100% transition to EV.

When we compared the current annual total emissions of all sectors using the EV scenario, it was revealed that carbon emissions will drop by about 41 $MtCO_2$ or by 11.6% annually.

25.4 Discussion

In this research, it did not cover the detailed life cycle assessment of CV and EV, as the focus of our study was the electricity consumption based on EV transition scenario and their effect on carbon emissions in the UK. The manufacturing of EV produces slightly higher amounts of carbon emissions than the CV. On the other hand, petrol/diesel cars require fuel, which first must be extracted, processed, and transported to gas stations, all of which generated additional carbon emissions. This is subject to change, as there are constant innovations, which would lead to a reduction in carbon emissions in manufacturing processes. Besides, lots of other variations would still need to be further explored such as energy use behaviour. For example, CV uses the engine's heat to warm up the interior space, while electric cars required electricity from the battery in order to provide thermal energy for the interior. These variables and associated future technology will have a significant impact on curbing the carbon emission within the transportation sector.

25.5 Conclusions

The expansion of electric cars in the UK would lead to an increase in electricity demand, which would result a rise in carbon emissions due to supply and manufacturing. On the other hand, transport emissions would drop significantly resulting in a total reduction of CO_2 levels by 12% per year. IT can be concluded that transition towards EV would gain a positive effect in reducing Co2 in the longer-term and further reduction can be achieved if energy efficiency increases at supply and manufacturing level.

References

1. Department for Business, Energy & Industrial Strategy, 2020. UK Becomes First Major Economy To Pass Net Zero Emissions Law. [online] GOV.UK. https://www.gov.uk/govern ment/news/uk-becomes-first-major-economy-to pass-net-zero-emissions-law. Accessed 25 May 2020
2. Web.archive.org. 2020. London Borough Of Bexley - Hiram Maxim And Edward Butler. [online]. https://web.archive.org/web/20160215204601, http://www.bexley.gov.uk/art icle/10664/Hiram-Maxim-and-Edward-Butler. Accessed 25 May 2020
3. T. Lewis. *Electric Vs. Fuel Cell Vehicles: 'Green' Auto Tech Explained.* [online] live-science.com (2020). https://www.livescience.com/49594-electric-fuel-cell-vehicles-explainer. html. Accessed 25 May 2020
4. D. Hiskey, In 1899 Ninety Percent of New York City's Taxi Cabs Were Electric Vehicles. [online] Today I Found Out (2011). https://www.todayifoundout.com/index.php/2011/04/in-1899-ninety-percent-of-new-york-citys-taxi-cabs-were-electric-vehicles/. Accessed 11 May 2020
5. UK Government; Department for Transport, 2018. Vehicle Licensing Statistics: Quarter 1 (Jan–Mar) 2018. Department for Transport, UK government, pp. 1–8
6. D. Moss, How do electric cars work?. [online] What Car? (2019). https://www.whatcar.com/advice/buying/how-do-electric-cars-work/n18091. Accessed 11 May 2020
7. K. Holmberg, A. Erdemir, The impact of tribology on energy use and CO_2 emission globally and in combustion engine and electric cars. Tribol. Int. **135**, 389–396 (2019)
8. UK Government; Department for Transport, *Provisional Road Traffic Estimates Great Britain: October 2018 - September 2019* (Department for Transport, UK Government, 2019), pp. 1–6
9. B. Lane, Electric Vehicle Market Statistics 2020 - How Many Electric Cars In UK &#63. [online] Nextgreencar.com (2020). https://www.nextgreencar.com/electric-cars/statis tics/. Accessed 25 May 2020
10. Department for Business, Energy & Industrial Strategy, 2019. Digest Of UK Energy Statistics (DUKES): Electricity. Digest of UK Energy Statistics (DUKES): annual data. UK Government, pp. 1–105

11. Statista, UK: Total Electricity Consumption 2002–2019 | Statista. [online] (2020). https://www.statista.com/statistics/322874/electricity-consumption-from-all-electricitysuppliers-in-the-united-kingdom/. Accessed 20 April 2020
12. GOV.UK, Energy Trends: UK Electricity. [online] (2020). https://www.gov.uk/government/statistics/electricity-section-5-energy-trends. Accessed 22 April 2020
13. Parliamentary Office of Science & Technology, 2011. Carbon Footprint Of Electricity Generation. Houses of Parliament, pp. 1–4
14. Data.gov.uk, Provisional UK Greenhouse Gas Emissions National Statistics - Data.Gov.Uk (2020)
15. UK Government, Electricity Distribution Losses. A consultation document. UK Government, pp. 1–30 (2003)

Chapter 26
The Effect of Temperature Variation on Bridges—A Literature Review

Sushmita Borah, Amin Al-Habaibeh, and Rolands Kromanis

Abstract Bridges are commonly subjected to complex load scenarios in their life-time. Understanding the response of bridges under such load scenarios is important to ensure their safety. While static and dynamic loads from vehicles and pedestrians influence the instantaneous response of bridges, studies show that thermal load from diurnal and seasonal temperature variation influences its long-term response and durability. This study addresses the effects of thermal load variation on bridges and briefly reviews methods of measuring such effects. The findings show that thermally induced deformations in bridges are of magnitude equal or larger than that induced by vehicle induced load. This study highlights the significance of measuring temperature responses of bridges for their robust structural health monitoring.

Keywords Bridges · Sensors · Structural health monitoring · Sustainable infrastructure · Thermal load

26.1 Introduction

Bridges are a vital element of road infrastructure. The serviceability of bridges is sensitive to continuous traffic loads and environmental impact due to variations of ambient temperature and wind loads. A robust structural health monitoring (SHM) approach is essential to determine bridges' serviceability and thereby ensure traffic safety. Morandi bridge collapse in 2018, which took the lives of 43 people, in Italy is a recent reminder of the need for robust SHM for bridges [1]. Figure 26.1 illustrates a general framework of monitoring bridges under common loading scenarios such

S. Borah (✉) · A. Al-Habaibeh
Product Innovation Centre, Nottingham Trent University, Nottingham, UK
e-mail: sushmita.borah2018@my.ntu.ac.uk

A. Al-Habaibeh
e-mail: amin.al-habaibeh@ntu.ac.uk

R. Kromanis
University of Twente, Enschede, The Netherlands
e-mail: r.kromanis@utwente.nl

© The Author(s) 2021
I. Mporas et al. (eds.), *Energy and Sustainable Futures*, Springer Proceedings in Energy,
https://doi.org/10.1007/978-3-030-63916-7_26

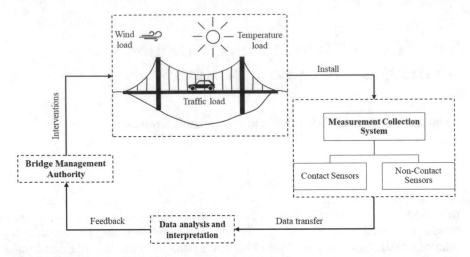

Fig. 26.1 A general framework of Bridge health monitoring

as traffic, temperature and wind load. Structural response data from the bridges are collected using suitable SHM approaches. Traditionally, in many countries general bridge inspection is carried out periodically that involves a visual survey of all accessible parts of the bridge. Such inspections are subjective and prone to human error. Contact and non-contact sensors are commonly used in SHM to eliminate the limitations of the general bridge inspection and conduct more frequent monitoring. The collected responses such as displacement, strain and accelerations are then analysed to provide feedback to bridge management authorities to undertake any required interventions.

The traffic load influences the instantaneous response of bridges, whereas, thermal load from diurnal and seasonal temperature variation influences its long-term response and durability. The magnitude of thermally induced response has been found in many cases to be equal or larger than the traffic-induced response [2]. This study aims to explore the literature on temperature-induced bridge responses and review key thermal response measurement techniques. The need for continuous SHM of bridges is emphasised to administer timely repair and replacement work. Early detection of damage can reduce the cost of bridge replacement and repair; and increase bridge life and traffic safety. Thus, SHM can aid sustainable infrastructure by ensuring public safety and reducing life-cycle maintenance cost of bridges.

26.2   Thermal Response of Bridges

26.2.1   Temperature Distribution Pattern in Bridges

Distribution of ambient temperature in bridges in most cases is non-linear. The temperature distribution is influenced by material type, bridge orientation, shading from neighbouring structures, etc. Kromanis and Kripakaran [3] recorded maximum temperatures from 26.8°C to 35.4 °C at different locations along the length of the National Physical Laboratory footbridge in Exeter, UK. In Shanghai Yangtze River Bridge, Zhou and Sun [4] observed a negligible temperature gradient in the steel girder along the longitudinal and transverse direction and a significant vertical temperature gradient between the top and bottom girder plates up to 17 °C on a hot summer day. Lawson et al. [5] computed temperature profiles along the depth of two hypothetical concrete and composite girders based on weather data of Nevada, USA. Daily temperature variation was maximum in the top surface of concrete superstructure followed by bottom and internal layers in decreasing order. In the composite superstructure, the temperature varied only in the concrete layer whereas the steel layer experienced almost uniform temperature. The temperature difference between any two points along the depth of both superstructures was found usually to be more than 30°C; as maximum temperature difference exceeded 40 °C in the concrete superstructure and 36°C in the composite superstructure. Thus, a significant temperature gradient is observed in the bridge due to ambient temperature and solar radiation, especially along its depth.

26.2.2   Temperature Effects on the Static Response of Bridges

Temperature gradients can induce static responses in bridges such as displacement and strain. Long term monitoring of Cleddau Bridge in Wales showed that bearing displacements closely follow diurnal and seasonal temperature variations [3]. Xia et al. [2] recorded daily temperature-induced strains of magnitude equal or larger than that due to traffic and static load in the Yangtze River in Jiangsu, China. Measured temperature-induced strain in the bridge exceeded 150 $\mu\varepsilon$ in the lower deck and 100 $\mu\varepsilon$ in the upper deck. In comparison, the traffic-induced strain in the bridge did not exceed 100 $\mu\varepsilon$, while strain under a point load of 17680 kN was found to be 156 $\mu\varepsilon$. Likewise, range of variation of thermal and vehicle induced stress in the lower deck were found to be of similar magnitude.

Zhou and Sun [6] monitored the Shanghai Yangtze River Bridge to evaluate temperature effects on several bridge response parameters. Temperature dependencies of these response parameters showed two modes: (i) amplitude of variations due to annual temperature cycles was significantly larger than those due to diurnal temperature cycles. Girder length, the distance between the two-tower tops and structural total strains at mid-span exhibited this mode and their temperature-induced change was governed by and directly proportional to average girder temperature. (ii) the amplitude of variation due to annual temperature cycles was approximately equal to those due to diurnal temperature cycles. The vertical deflection and elastic strains at mid-span, and average cable tensions of the longest centre-span cables exhibited this mode. The temperature-induced changes in mid-span vertical deflection and average cable tension were simultaneously governed by average cable temperature and average girder temperature. The elastic strain within the top and bottom plates of the girder was mainly caused by the temperature gradient between the plates.

26.2.3 Temperature Effects on the Dynamic Response of Bridges

Temperature gradient significantly influences the dynamic properties of bridges such as natural frequency. Cornwell et al. [7] reported about 5% variation of the first three natural frequencies of the Alamosa Canyon Bridge during daily temperature cycle. Liu and DeWolf [8] reported a maximum of 6% change in natural frequency of a concrete bridge in Connecticut, USA in a 21.11 °C peak to peak temperature change during one-year and found that changes in girder temperature affect the bending modes more easily than the torsional modes. Peeters & De Roeck [9] found a bilinear relationship between natural frequency and temperature change in Z24 Bridge. Natural frequency fluctuated between 1.7% to 6.7% over one-year in Ting Kau Bridge when temperature ranged between 3 °C and 53 °C [10]. The frequency of the first and fourth bending modes of the Tianjin Yonghe Bridge varied by 3.155% and 1.470% with a corresponding change in the ambient temperature of −11.5 °C to 3.7 °C over 16-days [11]. In Dowling Hall Footbridge, the first six modal frequencies varied by 4 to 8% within a temperature range of −14 °C to 39 °C over 16-weeks of data [12].

Several lab-based investigations also confirmed the dependency of the natural frequency of bridges with temperature variation. In a two-year dynamic test on a reinforced concrete slab in a laboratory, Xia et al. [13] observed a decrease of 0.13 to 0.23% in bending modes of natural frequency per one degree Celsius increase in temperature. Balmes et al. [14] observed an increase of 16%, 8%, 5%, and 3% in the first four natural frequencies of a clamped beam when ambient temperature decreased by 17 °C. Kim et al. [15] found a decrease of 0.64%, 0.33%, 0.44% and 0.22% in the first four natural frequencies of a small-scale laboratory bridge model per °C increase in temperature.

26.2.4   Health Monitoring Technology

The majority of the bridge monitoring events to measure thermal response used contact sensors that require physical deployment of hardware. The Yangtze River in Jiangsu, China was monitored with an SHM system of 170 sensors including accelerometers, Fibre Bragg grating (FBG) sensors to measure deck strain and temperature, displacement sensors, global positioning system (GPS) receivers, shear pins and ultrasonic anemometer, etc. [2]. Shanghai Yangtze River Bridge was mounted with 227 sensors including FBG and GPS [4]. Liu and DeWolf [8] studied thermally induced changes in natural frequency of a concrete bridge in Connecticut, USA that was equipped with 12 temperature sensors, 16 accelerometers, and six tiltmeters.

These contact sensors provide accurate measurement collection. However, their advantages are limited due to cumbersome and expensive installations and requirements of a dense sensor network [16]. Vision-based monitoring is an emerging non-contact measurement collection technique using cameras and image processing algorithms. Vision-based monitoring exhibited promising results in measuring traffic-induced responses [17]; however, limited studies are available measuring temperature-induced response [18]. Feasibility of such vision-based sensors in measuring thermal response can be explored in future.

26.3   Conclusions

This paper has presented a literature review of temperature variation effects in bridge responses such as displacement, strain and natural frequency and methods of measuring such responses. It concludes that temperature variations could have a slow but very significant effect on the long-term response of bridges, and it should be taken into consideration when addressing the design and monitoring of bridges. Contact sensors are commonly used in previous researches while emerging vision-based monitoring system can be explored in future for thermal response measurement. Future work will include developing a vision-based thermal response measurement collection technique and validating through finite element analysis and laboratory experiments.

References

1. G. J. O'Reilly et al., Once upon a Time in Italy: the Tale of the Morandi Bridge, Struct. Eng. Int., 1–20 (2018)
2. Q. Xia, L. Zhou, J. Zhang, Thermal performance analysis of a long-span suspension bridge with long-term monitoring data. J. Civ. Struct. Heal. Monit. **8**(4), 543–553 (2018)

3. R. Kromanis, P. Kripakaran, Predicting thermal response of bridges using regression models derived from measurement histories. Comput. Struct. **136**, 64–77 (2014)
4. Y. Zhou, L. Sun, Insights into temperature effects on structural deformation of a cable-stayed bridge based on structural health monitoring. Struct. Heal. Monit. **18**(3), 778–791 (2019)
5. L. Lawson, K.L. Ryan, I.G. Buckle, Bridge temperature profiles revisited: thermal analyses based on recent meteorological data from nevada. J. Bridg. Eng. **25**(1), 1–11 (2020)
6. Y. Zhou, L. Sun, A comprehensive study of the thermal response of a long-span cable-stayed bridge: from monitoring phenomena to underlying mechanisms. Mech. Syst. Signal Process. **124**, 330–348 (2019)
7. P. Cornwell, C.R. Farrar, S.W. Doebling, H. Sohn, Environmental variability of modal properties. Exp. Tech. **23**(6), 45–48 (1999)
8. C. Liu, J.T. DeWolf, Effect of temperature on modal variability for a curved concrete bridge. J. Struct. Eng. **133**(12), 1742–1751 (2007)
9. B. Peeters, G. De Roeck, One-year monitoring of the Z24-bridge: environmental effects versus damage events. Earthq. Eng. Struct. Dyn. **30**, 149–171 (2001)
10. H.F. Zhou, Y.Q. Ni, J.M. Ko, Constructing input to neural networks for modeling temperature-caused modal variability: mean temperatures, effective temperatures, and principal components of temperatures. Eng. Struct. **32**(6), 1747–1759 (2010)
11. H. Li, S. Li, J. Ou, H. Li, Modal identification of bridges under varying environmental conditions: temperature and wind effects. Struct. Control Heal. Monit. **17**, 495–512 (2010)
12. P. Moser, B. Moaveni, Environmental effects on the identified natural frequencies of the Dowling Hall Footbridge. Mech. Syst. Signal Process. **25**(7), 2336–2357 (2011)
13. Y. Xia, H. Hao, G. Zanardo, A. Deeks, Long term vibration monitoring of an RC slab: temperature and humidity effect. Eng. Struct. **28**(3), 441–452 (2006)
14. E. Balmes, M. Corus, D. Siegert, Modeling thermal effects on bridge dynamic responses, in: *Proceedings of the 24th International Modal Analysis Conference (IMAC-XXIV)* (2006)
15. J.T. Kim, J.H. Park, B.J. Lee, Vibration-based damage monitoring in model plate-girder bridges under uncertain temperature conditions. Eng. Struct. **29**(7), 1354–1365 (2007)
16. D. Feng, M. Q. Feng, Computer vision for SHM of civil infrastructure: from dynamic response measurement to damage detection—a review, Eng. Struct. **156**, 105–117 (2018)
17. Y. Xu, J.M.W. Brownjohn, Review of machine-vision based methodologies for displacement measurement in civil structures. J. Civ. Struct. Heal. Monit. **8**(1), 91–110 (2018)
18. R. Kromanis, Y. Xu, D. Lydon, J. Martinez del Rincon, A. Al-Habaibeh, Measuring structural deformations in the laboratory environment using smartphones, Front. Built Environ. **5** (2019)

Chapter 27
The Future of Hybrid Electric Vehicles and Sustainable Vehicles in the UK

Greg Last, David E. Agbro, and Abhishek Asthana

Abstract This paper details the development of the hybrid electric vehicle (HEV) and its integration into the UK market. The aim of this research was to explore the benefits and limitations of the HEV system which there are many. Government policies and incentives; both current and future as well as HEV technologies are also summarised. The HEV is an excellent short to medium term solution for making travel more sustainable. However, in the long term, push for electric vehicles (EVs) will significantly increase from the Government in its aim to meet stringent emissions policies and there will likely be legislation to phase out HEVs that cannot be plugged in.

Keywords Hybrid electric vehicles · Sustainable travel · CO_2 emissions · Electric vehicles · Government policies

27.1 Introduction

With the current climate crisis facing the world, there is a big push for countries to reduce their carbon emissions and lower the levels of air pollution affecting human health. In the UK, a large contributor to air pollution and climate change is the transport sector, responsible for around 24% of total UK emissions [1]. Transportation includes travel by Air, Water and Road, with road transportation accounting for 72% of the CO_2 emissions. The automobile makes up the largest proportion of the road transport, at around 61% [2] and holds the opportunity to significantly improve the UK's total carbon emission. Reducing tailpipe emissions is hugely important for the planet and human health.

This report aims to provide useful and current information on the status of HEVs in the UK, specifically the benefits and pitfalls from both a performance and environmental angle. Government policies and incentives; both current and future as well as HEV technologies are outlined. The report will make for an easier understanding

G. Last · D. E. Agbro (✉) · A. Asthana
Sheffield Hallam University, Howard St, Sheffield S1 1WB, UK
e-mail: d.agbro@shu.ac.uk

© The Author(s) 2021
I. Mporas et al. (eds.), *Energy and Sustainable Futures*, Springer Proceedings in Energy,
https://doi.org/10.1007/978-3-030-63916-7_27

Fig. 27.1 CO₂ Emission contributions from each transport mode in the UK as of 2016 [2]

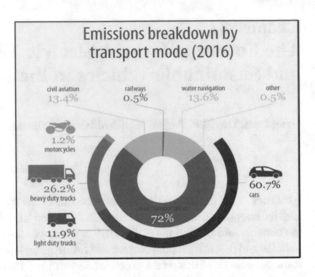

of all aspects of owning an HEV in the UK, whereby an informed decision can be made on whether an HEV is suitable for the individual, or not (Fig. 27.1).

27.2   Mainstream Integration of HEV's in UK

The first popular HEV to hit a consumer market was the Toyota Prius in Japan, in 1997. 18,000 were sold in the first year and two years later, the Honda Insight was launched, available in Japan and the US. The Prius shortly followed and was lauched in the US, Europe and the rest of the world in 2000 [3]. These two cars marked the start of the large scale production of HEV's across the Globe. In 1999 Honda sold around 200 Insight's making the model the first actual HEV sold commercially in the UK [4]. With such low number of Honda Insight's bought, the Toyota Prius has become generally accepted as the first HEV on the UK market in 2000 and became the world's top selling HEV—in 2017 Toyota had sold 4 million units globally [3]. The Prius marked the start of commercial HEV's as a viable choice available for the consumer in the UK. The economic advantages of a hybrid became instantly obvious to the more frugal consumer. With the initial version of the Toyota Prius, owners could expect a combined 60 mpg compared to the Toyota Corolla, a similar 4 door petrol Internal Combustion Engine (ICE) vehicle with an average combined 31mpg [5]. This comparison is true for most ICE—HE vehicles with the fuel economy been far better in the latter. Over the last two decades, the HEV has developed with more and more manufacturers offering a HEV in their range. Many of these models have been available in the UK market, however the Prius has remained at the top of the list with the UK being one of the leading EU markets for the model since 2000, taking 20% of Prius sales by 2010 [6]. Although the UK has been a seemingly big advocate of the HEV, the market share of HEV's across the total automotive market

Fig. 27.2 Market share of non-ICE automobiles in the UK from 2010–2015 [8]

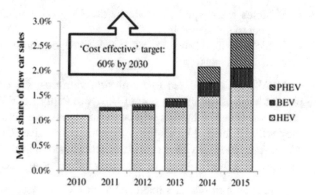

has been quite small by proportion; in 2010 just 1.1% of all automobile sales were HEV's and by 2015 that had only risen to 1.6% [7]. Figure 27.2 shows the shares of non-ICE automobiles in the UK including plug-in hybrid electric vehicle (PHEV), battery electric vehicle (BEV) and hybrid electric vehicles (HEV).

However, the market share is on the rise, as is the demand for HEV's, and as of October 2019, 10% of the market is now held by alternate fuel vehicles, including 5.5% HEV's [9]. This shows that both consumer and manufacturer are making a step in the right direction for sustainable travel. For the consumer, the environmental impact of driving an ICE vehicle is a consideration but with the cost of fuel rising and the economic climate in the UK still unsettled, the demand for lower fuel consumption is a big driver of sales of HEV's. So, the HEV like the Toyota Prius, or similar offerings from Honda, Ford, VW, etc. are great for those who see automobiles as a mode of travel from A–B only. Most of the HEVs sold on the market over the last two decades have left much to be desired in aspects of aesthetics and performance. The 2011 Toyota Auris for example was built in Burnaston, UK [10]. A great HEV offering and built in the UK, this makes it a sustainable choice, yet from a performance vehicle aspect it simply falls short.

27.3 Government Driven Policy

27.3.1 HEV Taxation

For HEV's registered from April 2017 onwards, the British Government has issued an Alternative Fuel Discount (AFD) which is a £10 reduction of the First Year Rate (FYR) and the Standard Rate each year after. For new HEV's the greatest benefit comes in the first year in the form of the FYR. The FYR is calculated on tailpipe CO_2 emissions, an area where HEV's perform better than ICE vehicles. The true beneficiaries of tax rate savings are the vehicles with emission output of less than 75 g/km.

- Vehicles rates equal to or are below 75 g/km are £25 or less, decreasing with respect to emissions levels.
- Vehicles rates above 75 g/km are £110 or more, increasing with respect to emissions levels.

HEV's are considered low emission vehicles, yet only one model on the new market has a 75 g/km CO_2 output—Toyota Prius 1.8 VVT-I Active Auto [11]. This means the other 29 models will be in the higher cost bracket for the FYR, although they will still be 3–4 bands lower than the respective ICE vehicles and have the £10 AFD. After the first year, the tax bracket is divided into Standard and Premium, with the latter bracket reserved for vehicles costing >£40,000.

- The Premium rate is a fixed £465 for CO_2 emissions above 50 g/km, which all new HEV's are.
- The Standard rate (for <£40,000 vehicles) is a fixed £150 [17] for CO_2 emissions above 50 g/km, which all new HEV's are.
- The £10 AFD does apply to both the Standard and Premium rates for HEV's. So, there is a £10 saving per annum for HEV's over ICE vehicles after the first year also.

The greatest savings, therefore, are through HEV's registered between March 2001 to March 2017. Alternative Fuel Vehicles (AVF), of which an HEV is, receive a £10 reduction, and since the tax rate is only calculated by vehicle CO_2 emissions, the HEV's generally sit 3–4 VED bands lower than their ICE counterparts, generating a substantial saving each year. Further to this, the lowest VED band is from CO_2 emissions of up to 100 g/km, which many HEV's meet, and the AFV rate is £0. As shown in Fig. 27.3, there is a considerable hike in cost between VED Band C and D, since many HEV's fall into Band C or lower, the savings are considerable over the ICE vehicles which are often Band D or higher [11].

12-month VED rates for cars registered from 01 March 2001 to 31 March 2017*			
VED Band	Vehicle CO2 emissions	Standard rate 2019-20	AFV rate** 2019-20
A	Up to 100 g/km	£0	£0
B	101-110 g/km	£20	£10
C	111-120 g/km	£30	£20
D	121-130 g/km	£125	£115
E	131-140 g/km	£145	£135
F	141-150 g/km	£160	£150
G	151-165 g/km	£200	£190
H	166-175 g/km	£235	£225
I	176-185 g/km	£260	£250
J	186-200 g/km	£300	£290
K*	201-225 g/km	£325	£315
L	226-255 g/km	£555	£545
M	Over 255 g/km	£570	£565

Fig. 27.3 The current VED rates for Mar '01–Mar '17 vehicles [11]

27.3.2   Alternative Fuel Vehicle Initiatives

Until October 2018, there was a Plug-In grant for PHEV's and EV whereby a subsidy was available for buyers of these vehicles. Although not available to HEV's due to their very limited range in Zero-emissions mode (Electric Only), this initiative showed that the British Government supported Hybrid-Electric vehicles as a sustainable transport method. However as of October 2018, the grant was changed to support only EV's, with no PHEV support. This decision came about because research suggested that many of the PHEV's were not actually being plugged in and charged. The Government saw this as undermining the incentive and the environmental benefits [12]. Essentially, a PHEV that is not plugged in, acts just like a HEV, and since the Government is no longer supporting the PHEV because of this—the support for HEV could be questioned.

27.3.3   Green Initiatives

In 2008 the UK Government brought in the Climate Change Act. Along term, legally binding climate change reduction legislation. The Act laid a framework for emissions reductions and established five-year 'carbon budgets' up to 2050 [13]. Through this, the UK became the first major economy across the globe to pledge the reduction of its contribution to climate change and global warming by 2050 in a legally binding contract. In 2015, the UK entered into The Paris Agreement, a policy spearheaded by the United Nations Framework Convention on Climate Change (UNFCC). This was truly a global initiative developed to hold the increase in global average temperature to below 2 °C above pre-industrial temperatures and aim to limit global warming to just 1.5 °C. Signed by 160 countries, this marked the turn in environmental awareness and impact. From this, the UK agreed on a target of 40% emissions reductions from 1990 levels, by 2030 [13]. As of 2019, the UK surpassed the target of the Paris Agreement, with a 42% reduction in emissions, 11 years ahead of schedule. This led to the decision to revise the Climate Change Act of 2008, and on 27th June 2019, the UK have pledged to end it's contribution to climate change by 2050. The legally binding contract lays out a target that by 2050, the UK's net emissions will be zero [14].

For the UK to meet this target, the transport sector and respectively the automotive sector will need to be zero-emissions come 2050. Simply stated, this cannot be met through the use of HEV's or PHEV's. Both hybrid types have an onboard ICE and therefore produce tailpipe emissions. This is a topic of debate that is currently ongoing in government, specifically amongst the Business, Energy and Industrial Strategy Committee (BEISC) and the Committee on Climate Change (CCC). Stated in the Electric Vehicles: driving the transition (fourteenth Report of Session 2017-2019) document provided to the House of Commons, the CCC advised that only BEV and long-range PHEV's should be eligible for sale 2035 onwards [15]. The idea is that the majority of all vehicle trips can be completed without an ICE. There is quite some inconsistency reported in the document however as to whether the HEV will be covered by the 'conventional' vehicle phase out or not, since it cannot be plugged in. The 'conventional' vehicle phase out requires the ban of these vehicles by 2040, which if HEV's are to be considered in this group, would spell the demise of the HEV within the next 20 years. Ultimately, there is still a very ambiguous and unclear decision on whether to phase out HEV's or not, but for the Government to meet it's Zero-Emissions target, the HEV without plug in option is not compatible with this commitment. If a ban on HEV's is not imposed by 2040, the target of only allowing the cleanest new vehicles to the market at that time will be undermined [15].

27.4 What Is the Future for the HEV?

During the phase out of the 'conventional' vehicle, HEV included or not, the HEV and PHEV will play an integral part as a stop gap between the removal of pure ICE cars and the dominion of the EV. Many manufacturers are committed to the removal of ICE vehicles from the roads yet support the Hybrid platform as a means of long-term emissions reductions—albeit not cutting emissions to zero. It is clear that for the Government to meet the 2050 target of net zero-emissions, the HEV will not be a long-term solution. However, with very limited infrastructure across the UK to support EV's, the HEV holds an important role of providing a more sustainable option over traditional ICE vehicles. Until the infrastructure is in place for EV's, HEVs and PHEVs can provide transitional technology that alleviates the common 'range anxiety' of prospective and/or current EV owners. Infrastructure is not the only part that will improve for the EV, battery technology is progressing fast, allowing increasing ranges of EV's. The pressure from Governments like the UK's, forces manufacturers to invest in the research and development, benefiting HEVs, PHEVs and EVs [15].

27.4.1 The Effect of Brexit

Leaving the EU would result in a change to how the carbon budget in the UK is delivered. With severed ties, any previous agreements on policy will no longer apply or at the very least will be severely weakened. The UK Government will need to replace these with policies of their own since the needs to tackle global warming does not waiver [13]. The UK is world leading in the strive for net zero emissions, and although Brexit looms overhead, the result should not ultimately affect the UK's path towards this goal. However, for the consumer, Brexit does leave a continued uncertainty around exiting the EU and this has affected vehicle sales. There is a growth in Hybrid and Electric vehicle sales, which shows consumers are keen on the new technology, however, a lack of consumer confidence caused by Brexit leaves the automotive market soft. In October 2019, a 6.7% fall in new car sales was seen compared to 12 months prior, with the general election, due 12th December and as 2019 draws to a close, this is unlikely to improve [9].

27.4.2   Issues and Limitations

Although HEV's offer a serious potential for consumers to decrease their carbon footprint and environmental impact, there is a fundamental flaw. Many of the HEV's rely on regenerative braking to recharge the batteries, unlike PHEV's with the option to charge off the mains, the HEV requires journeys with a lot of deceleration. In urban driving, this is often true, with a lot stop-start in traffic. However, on long haul motorway journeys, the braking is reduced considerably and often the vehicle does not fully recharge its batteries. The result is that high potential fuel savings reduce to the levels of ordinary ICE powered vehicles, a major consideration for prospective buyers [4].

As the UK's environmental awareness improves, so does the number of AFVs on the road. The servicing and maintenance network for HEV's is therefore diversifying and moving away from only franchised dealers and manufacturers. However, there are still limited capabilities across the UK, and this is a factor to be aware of for prospective buyers. With the new hybrid systems, far higher voltages are present, and offer a serious threat of injury for untrained technicians. This means that not only is the technology different and the garages may not be familiar with the various systems, but a higher danger level is present, and some garages may not be comfortable with these vehicles [16]. Training and specialist equipment are therefore necessary and it's important that any garage the customer takes their HEV to is competent to carry out the required work—this could be a limiting factor for prospective buyers of second-hand HEV's who aren't covered by the dealership and haven't access to a suitable maintenance provider.

27.5   Conclusion

The current government policies do not really support the HEV as a sustainable mode of transport and the Tax savings are limited. It is clear than by 2040 at least, the HEV without plug in ability will be phased out. Although, no specific decision has been made on this, for the UK government to meet it's emissions target, this will have to be the case. There is however 20 years between now and 2040, a long time in respect to vehicle ownership. It is important that everyone tries where they can to reduce their impact on the environment, and with the current HEV models, although limited, there are some very good vehicles. The HEV is not future proof, nor is it wholesomely supported by any UK Government initiatives, however the

HEV provides huge fuel savings and low emissions output. This makes them a very real option, and for the enthusiast, the performance of the electric motor is hard to beat.

References

1. BEIS, Clean Growth Strategy: executive summary. Retrieved from GOV.UK (2018)
2. EEA, Co2 emissions from cars: facts and figures (infographics). Retrieved from European Parliament (2019)
3. A. Smith, D. Moss, A brief history of hybrid and electric vehicles—picture special. Retrieved from Autocar (2013). https://www.autocar.co.uk/car-news/frankfurt-motor-show/brief-history-hybrid-and-electric-vehicles-picture-special
4. R. Dredge, Hybrid cars explained. Retrieved from hpi (2017). https://www.hpi.co.uk/content/electric-cars-the-electric-era/hybrid-cars-explained/
5. EPA, Fuel Economy Compare. Retrieved from Fueleconomy.gov (2019), https://www.fueleconomy.gov/feg/noframes/18605.shtml
6. E. Loveday, Toyota Prius eclipses 200,000 total sales in Europe, doubling mark set in 2008. Autoblog Green, 50–52 (2010)
7. ICCT, European Vehicle Market Statistics. EU Pocketbook, 81–107 (2016)
8. C. Brand, C. Cluzel, J.L. Anable, Modeling the uptake of plug-in vehicles in a heterogeneous car market using a consumer segmentation approach. Transp. Res. Part A Policy Practice **97**, 121–136 (2017)
9. P. Campbell, Electric and hybrid car sales jump to 10 per cent of UK total. London, London, United Kingdom (2019)
10. just-auto.com, UK: Toyota prepares for hybrid production. Retrieved from just-auto.com (2010). https://www.just-auto.com/news/toyota-prepares-for-hybrid-production_id104353.aspx
11. C. Lilly, Car Tax Bands 2019/20. Retrieved from Next greencar (2018). https://www.nextgreencar.com/car-tax/bands/#q2
12. J. Holder, Exclusive: Government won't reinstate plug-in hybrid grants. Retrieved from Autocar.co.uk (2019). https://www.autocar.co.uk/car-news/industry/exclusive-government-wont-reinstate-plug-hybrid-grants
13. CCC, Legal duties on climate change. Retrieved from Committe on Climate Change (2018). https://www.theccc.org.uk/tackling-climate-change/the-legal-landscape/
14. DBEIS, UK becomes first major economy to pass net zero emissions law. Retrieved from gov.uk (2019), https://www.gov.uk/government/news/uk-becomes-first-major-economy-to-pass-net-zero-emissions-law
15. BEISC, *Electric Vehicles: driving the transition* (House of Commons, London, 2018)
16. HSE, Electric and hybrid vehicles. Retrieved from HSE.gov.uk (2019). http://www.hse.gov.uk/mvr/topics/electric-hybrid.htm
17. Vehicle Tax Rates (2020, 08 21). https://www.gov.uk/vehicle-tax-rate-tables

Part V
Energy Governance, Policy, and Sustainability

Chapter 28
First Step Towards a System Dynamic Sustainability Assessment Model for Urban Energy Transition

Bjarnhedinn Gudlaugsson, Huda Dawood, Gobind Pillai, and Michael Short

Abstract This paper presents a conceptual model that describes the correlation between an urban energy system and sustainability. The model captures the complexity of the urban energy transition, and the task of achieving sustainable development needs to embrace all aspects of sustainability. This paper portrays the aspects of sustainability as four-dimensional—Environment, Economic, Society, and Technology. The relationship between these four dimensions and the urban energy system is presented in a simplified and aggregated-qualitative based causal-loop diagram. The causal-loop diagram illustrates the causal and interconnective relationships between the four dimensions and their different variables. The causal-loop diagram describes the complex dynamic relationships within a simple urban energy system. The paper also provides a brief description of balancing and reinforcing loops, with the causal-loop diagram present. The conceptual model along with the causal-loop diagrams visually illustrate the dynamic relationship between the four dimensions as well as highlights the complexity and challenging problems that decision-makers are facing today when it comes energy planning and energy system development.

Keywords Energy system · System dynamics · Causal loop diagram · Decision-Making · Sustainability

28.1 Introduction

Today's cites account for 67% to 76% of the global energy consumption and as well as 71% to 76% of the global greenhouse gas (GHG) emission [1]. Majority of energy systems today operate on fossil fuels, fossil fuel resources account for approximately 86% of the total primary energy sources (TPES) in the global energy system. In contrast, renewable energy resources account for approximately 14% of the TPES [2, 3]. Global and local energy systems continue to grow and have become more

B. Gudlaugsson (✉) · H. Dawood · G. Pillai · M. Short
School of Computing, Engineering and Digital Technologies, Teesside University,
Stephenson Building, Stephenson Street, Tees Valley, Middlesbrough TS1 3BA, UK
e-mail: b.gudlaugsson@tees.ac.uk

© The Author(s) 2021
I. Mporas et al. (eds.), *Energy and Sustainable Futures*, Springer Proceedings in Energy,
https://doi.org/10.1007/978-3-030-63916-7_28

complex over recent years, due to factors such as changes in technology, energy source availability, energy regulation and policies; the environmental impacts of energy development and production have also increased, rising global environmental concerns [4]. Since urban energy systems are still primarily built around using fossil fuel resources for providing electricity, district heating, cooking, as well as public transportation, the global community can still expect to see a growing fossil fuel demand in coming years correlated with an increase in GHG emission [5, 6]. As global trends predict an increased growth in urbanisation, movement and relocation of people from the rural areas into larger community settlements like cities, there is a need to address sustainability issues of energy development, and transition of cities [1].

To achieve an energy transition towards a decarbonised energy system will be challenging as our current social development structure is primarily measured by economic growth, with environmental concerns relegated insignificance [7–9]. In particular, unchecked economic growth may increase the constraints on non-renewable and renewable natural resources and materials in our environmental sphere [10, 11]. In addition, is today's decision-making process, the decision-makers are often affected by "the silo effect", in which a lack of communication between subgroups within an organisation often results in a lack of cooperative decision making [12]. A redesign of the current decision-making process is needed to provide interconnectivity between societal, economic and environmental spheres [13–15].

Modelling and simulation approaches such as Multi-Criteria Decision Analysis, Life-Cycle Assessment and System Dynamics have become widely used to investigate complex issues such as sustainability, climate change and energy transitions to improve decision-making and policy strategies [12, 16, 17]. This paper will provide insight into the early development phase of designing a sustainability assessment model base on system thinking and system dynamics modelling to assess the sustainability of urban energy transition. This paper and its findings aim as well to present insights into the complex and dynamic interlinking relationships between the different aspects of sustainability and energy transition.

28.2 Methods

System dynamics is a method that can be classified as an interdisciplinary, and it applies the theory of system thinking and system structure to investigate complex systems. Researchers have been applying system dynamic modelling to address, understand and define complex and dynamic behaviour, feedback mechanisms, multidimensional aspects and causal relationship of a complex system [16, 18, 19]. A complex system is defined by Mitchell 2009 (p.13) as "a system in which larger networks of components within no central control and simple rules of operation give rise to complex collective behaviour, sophisticated information processing, and adaptation via learning or evolution" [20]. System dynamic analysis is built around using balancing and reinforcing feedback loops to represent the causal relationship, dynamic feedback and flow pathways within a system in order to identify the dynamics which arise out of these interactions [18, 21]. A Causal loop diagram (CLD) is a tool used in the early stage in system dynamic modelling process to gain visual understanding connection, through using the basic elements of words, phrases, links and loops. By drawing a simple CLD of the problem and the variables within the system become easier to understand the balancing and reinforcing feedback loops and the relationship between different variables, which thereby helps to identify the positive and negative relationships and any potential time delays within the system [22]. A balancing loop describes a relationship that seeks to keep the balance and a stable condition of the system when a change arises by counteracting the effect that leads to balance in the system [21–23]. In comparison, a reinforcing loop seeks to amplify and reinforce changes in the system, which can often lead to exponential growth, which can also leave a negative impact on the whole system [21–23]. The next step is to apply mathematical equations, and computer modelling approaches to describe these feedback loops to be better able to capture and analyse the dynamic elements of this complex system [24–26].

28.3   Results and Discussion

Figure 28.1. presents the conceptual framework of the problem and where the problem of the complexity of achieving a sustainable energy transition is defined in the centre of the conceptual model. The boundaries of the system are defined by the dot-line circle around the four dimensions—Environment, Economic, Society, Technology. These four dimensions represent the metrics sets which are often used and quantified when conducting a sustainability assessment [10, 12, 15, 16].

28.4   System Dynamic Model—Causal-Loop Diagram and Feedback Mechanism

The CLD presented in this paper is a qualitative based model, based on extensive literature review, and derives some of its causal relationships and feedback mechanism from the literature [16, 19, 23]. Figure 28.2 presents a large-scale and abstract CLD view of the evolving system dynamic model and the model's balancing loops and reinforcing loops. This CLD shows the groundwork for a system dynamic model that can conduct a sustainability assessment of energy transition within an urban energy system. The full development of that model will take place in the next steps of the research project.

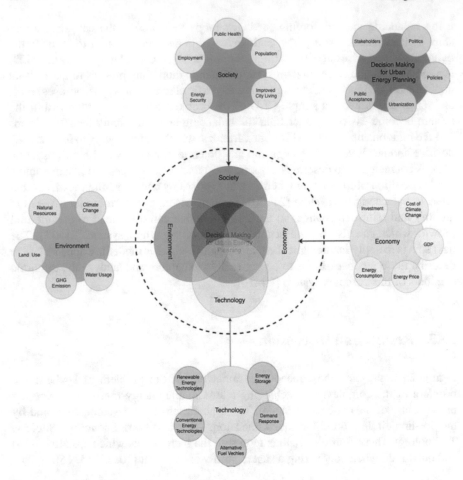

Fig. 28.1 Conceptual model and boundaries of the modelling system

The balancing loops in Fig. 28.2. illustrate the goal-seeking or stability-seeking causal relationships between the different variable in the model. This causal relationship of the balancing loops is described in the following paragraphs. Balancing loops (3) and (4) show that an increase in energy production will result in an acceleration of overall climate change and the ecological impacts, causing an increase in economic damage of climate change, thereby affecting the investments in energy system development and energy production capabilities.

The reinforcing loops in Fig. 28.2. illustrate the amplifying or self-multiplying causal relationships between the different variables in the model. This causal relationship of the reinforcing loops is described in the following paragraphs. Reinforcing loop (2), population growth increases the number of households requiring energy and energy consumption which grows the energy demand. Growing energy demand,

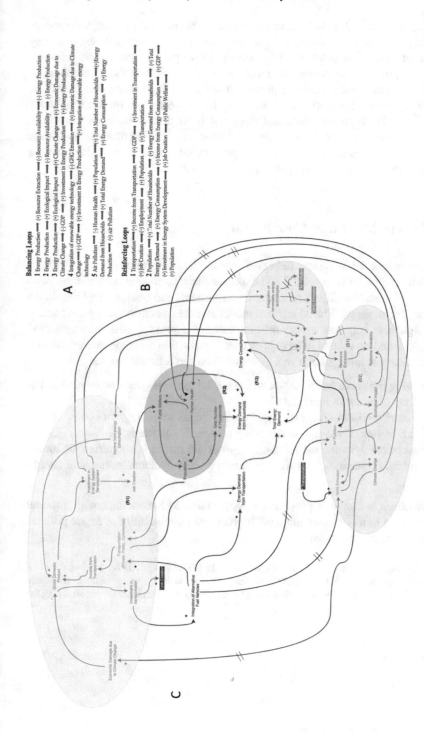

Fig. 28.2 **a** Balancing Loops, **b** Reinforcing Loops, and **c** Large-Scale View Causal Loop Diagram of a first draft of a Sustainability Assessment of Urban Energy Transition

leads to higher income for the local and national government from energy consumption, giving the local and national governments the opportunities to invest in energy system development that provides an increase in job opportunities, and employment among the population and which increases living quality among the populations.

The delays within the system are presented in Fig. 28.2. by using delays marker, which are represented with two parallel lines. These delays interpret a time delay where a cause leads to an effect that does not immediately have an impact on the system. For example; resource extraction for energy production will over time decrease the resources available for energy production, which then will impact the system abilities to meet the energy demand of the system.

28.5 Future Steps and Conclusion

The conceptual model illustrated in Fig. 28.1 is the basis of the causal loop diagram presented in this paper in Fig. 28.2 and captures the relationship between the three dimensions of sustainability (Economic, Environment, and Society) and the fourth dimension (Technology). The conceptual model and the causal-loop diagram capture and highlight the complex and multi-dimensional aspects of the interconnected relationships existing related to an energy transition within an urban energy system.

The next step in this research will be focused on the following tasks:

- Quantifying the variables presented in the causal loop diagram and expand on the causal loop diagram in more details.
- Make a stock-and-flow diagram based on the causal loop diagram
- Develop a System Dynamic Model using computer software
- Test the system dynamic model using appropriate methods and refine the model to ensure its robustness for assessing urban energy systems sustainable energy transition.

This research is aimed at providing policymakers and academics with a model that can be used to test and simulate the impact and sustainability of future policies in relation to the energy transition.

Acknowledgements The authors would like to thank the DTA3/COFUND for funding the PhD, through the European Union's Horizon 2020 research and innovation programme under the Marie Sklodowska-Curie grant agreement No 801604. Special thanks to Lilian Chiu for proofreading the paper.

References

1. United Nations, Department of Economic and Social Affairs, Population Division (2019). World Urbanization Prospects: The 2018 Revision (ST/ESA/SER.A/420). New York: United Nations. Accessed May 2020
2. K.E.H. Jenkins, D. Hopkins, *Transitions in energy efficiency and demand : the emergence, diffusion and impact of low-carbon innovation.* Routledge (2018). https://doi.org/10.4324/978 1351127264
3. IEA, World Energy Balances 2019, IEA, Paris (2019). https://www.iea.org/reports/world-ene rgy-balances-2019. Accessed May 2020
4. M. Mutingi, C. Mbohwa, V.P. Kommula, System dynamics approaches to energy policy modelling and simulation. Energy Procedia **141**, 532–539. Elsevier Ltd (2017). https://doi. org/10.1016/j.egypro.2017.11.071
5. J. Keirstead, M. Jennings, A. Sivakumar, (2012, August). A review of urban energy system models: approaches, challenges and opportunities. Renew. Sustain. Energy Rev. https://doi. org/10.1016/j.rser.2012.02.047
6. J. Rosales Carreón, E. Worrell, Urban energy systems within the transition to sustainable development. A research agenda for urban metabolism. Resour. Conserv. Recycl. **132**, 258–266 (2018). https://doi.org/10.1016/j.resconrec.2017.08.004
7. R. Costanza, H.E. Daly, Natural capital and sustainable development. Conserv. Biol. **6**(1), 37–46 (1992)
8. J. Harris, T. Wise, K. Gallagher, N. Goodwin (eds.), Basic principle of Sustianable development. A Survey of Sustainable Development: Social and Economic Dimensions, Volume 6 in the series Frontier Issues in Economic Thought (Washington, D.C.: Island Press, 2001)
9. IPCC, 2014: Climate Change 2014: Mitigation of Climate Change. Contribution of Working Group III to the Fifth Assessment Report of the Intergovernmental Panel on Climate Change [Edenhofer, O., R. Pichs-Madruga, Y. Sokona, E. Farahani, S. Kadner, K. Seyboth, A. Adler, I. Baum, S. Brunner, P. Eickemeier, B. Kriemann, J. Savolainen, S. Schlömer, C. von Stechow, T. Zwickel and J.C. Minx (eds.)]. Cambridge University Press, Cambridge, United Kingdom and New York, NY, USA
10. S. Luthra, S.K. Mangla, R.K. Kharb, Sustainable assessment in energy planning and management in Indian perspective. Renew. Sustain. Energy Rev. **47**, 58–73 (2015)
11. S. Moslehi, R. Arababadi, Sustainability assessment of complex energy systems using life cycle approach-case study: arizona state university tempe campus. Procedia Eng. **145**, 1096–1103 (2016). https://doi.org/10.1016/j.proeng.2016.04.142
12. N.C. Onat, M. Kucukvar, O. Tatari, Q.P. Zheng, Combined application of multi-criteria optimization and life-cycle sustainability assessment for optimal distribution of alternative passenger cars in U.S. J. Clean. Prod. **112**, 291–307 (2016). https://doi.org/10.1016/j.jclepro. 2015.09.021
13. K. Narula, B.S. Reddy, Three blind men and an elephant: the case of energy indices to measure energy security and energy sustainability. Energy **80**, 148–158 (2015). https://doi.org/10.1016/ j.energy.2014.11.055
14. I. Dincer, C. Acar, Smart energy systems for a sustainable future. Appl. Energy **194**, 225–235 (2017)
15. A. Bhardwaj, M. Joshi, R. Khosla, N.K. Dubash, More priorities, more problems? Decision-making with multiple energy, development and climate objectives. Energy Res. Soc. Sci. **49**, 143–157 (2019)
16. J.K. Musango, A.C. Brent, B. Amigun, L. Pretorius, H. Müller, A system dynamics approach to technology sustainability assessment: the case of biodiesel developments in South Africa. Technovation **32**(11), 639–651 (2012)
17. P.O. Siebers, Z.E. Lim, G.P. Figueredo, J. Hey, An innovative approach to multi-method integrated assessment modelling of global climate change. J. Artif. Soc. Soc. Simul. **23**(1), 1–10 (2020)

18. H. Zhang, J. Calvo-Amodio, K.R. Haapala, A conceptual model for assisting sustainable manu-facturing through system dynamics. J. Manuf. Syst. **32**(4), 543–549 (2013). https://doi.org/10.1016/j.jmsy.2013.05.007
19. N.C. Onat, M. Kucukvar, O. Tatari, Uncertainty-embedded dynamic life cycle sustainability assessment framework: an ex-ante perspective on the impacts of alternative vehicle options. Energy **112**, 715–728 (2016). https://doi.org/10.1016/j.energy.2016.06.129
20. M. Mitchell, *Complexity: A Guided Tour.* Oxford University Press (2009)
21. D.H. Meadows, *Thinking in Systems: A Primer.* Chelsea Green Publishing (2008)
22. J.D. Morecroft, *Strategic Modelling and Business Dynamics: A Feedback Systems Approach.* Wiley (2015)
23. N.C. Onat, M. Kucukvar, O. Tatari, G. Egilmez, Integration of system dynamics approach toward deepening and broadening the life cycle sustainability assessment framework: a case for electric vehicles. Int. J. Life Cycle Assess. **21**(7), 1009–1034 (2016). https://doi.org/10.1007/s11367-016-1070-4
24. H. Sverdrup, K.V. Ragnarsdóttir, Natural Resources in a Planetary Perspective. (E. H. Oelkers, Ed.), *Geochemical Perspectives. European Association of Geochemistry* (2014). https://doi.org/10.7185/geochempersp.3.2
25. B.K. Bala, F.M. Arshad, K.M. Noh, System Dynamics. Springer Texts in Business and Economics (2017)
26. T. Craig, A. Brent, F. Duvenhage, F. Dinter, Systems approach to concentrated solar power (CSP) technology adoption in South Africa. *In AIP Conference Proceedings* (Vol. 2033). American Institute of Physics Inc (2018). https://doi.org/10.1063/1.5067166

Chapter 29
How Often Do You Open Your House Windows When Heating is ON? An Investigation of the Impact of Occupants' Behaviour on Energy Efficiency of Residential Buildings

Sherna Salim and Amin Al-Habaibeh

Abstract Currently, there are many initiatives to thermally insulate buildings on the assumption that the more insulated the building is, the more efficient in terms of energy conservation it will perform. Many assessment systems assume a linear relationship between building insulation and energy conservation. The drawback of such hypotheses is that they ignore the effect of occupants' behaviour in their conclusions. In this study, the authors will examine the effect of people's behaviour, particularly windows' opening, as a behavioural pattern of occupants. It aims to study the impact of occupant's behaviour on energy consumption of residential buildings and to identify the key factors that influence occupants' behaviour; thus, providing ideas for improving energy efficiency by suggesting enhanced policies, approaches and techniques. The findings suggest that occupants' behaviour could have a greater influence on the energy efficiency of buildings in some cases when compared with their thermal insulation due to opening of windows in cold weather which causes air infiltration.

Keywords Energy efficient buildings · Energy consumption in buildings · Occupant behaviour · Air quality · Sustainability

29.1 Introduction

The UK government has committed to reduce its carbon footprint to nearly zero by the year 2050 [1]. To achieve this, a massive investment will be needed in clean energy generation and reductions in fossil fuel consumption. In this context several initiatives have been planned or taken place, such as the Green Deal, for insulating

S. Salim (✉) · A. Al-Habaibeh
Product Innovation Centre, Product Design, Nottingham Trent University, Nottingham, UK
e-mail: sherna.salim2017@my.ntu.ac.uk

A. Al-Habaibeh
e-mail: Amin.Al-Habaibeh@ntu.ac.uk

© The Author(s) 2021
I. Mporas et al. (eds.), *Energy and Sustainable Futures*, Springer Proceedings in Energy,
https://doi.org/10.1007/978-3-030-63916-7_29

1 to 3.7 million houses with solid wall insulation, by 2030. There has been a steady increase in the number of households in the UK, since 1991, contributed by factors such as the increase in birth rate, net immigration and the long-term trend of single adult households [2]. According to International Energy Agency, energy use in buildings is influenced by six parameters: climate, building envelope, building energy and services system, indoor design criteria, building operation and maintenance and occupants' behaviour [3, 4]. Providing smart meters to every home and business by 2020 is another government commitment. Several policies have been brought forward by the government in the past decade to improve energy efficiency in the domestic sector. The green deal, The Energy Act 2011, incentives to improve insulation in houses (solid wall insulation and loft insulation) all concentrate on improving building performance by improving the fabrics of existing buildings. However, government statistics show that despite improvement to the buildings' fabric, houses do not meet the original energy set targets. This has been inferred as a result of a report after the analysis of data from a subset of 76 homes by the Innovate UK's Building Performance Evaluation Programme (BPE) [5]. In its strategies to achieve carbon budgets, it has been clearly stated that '*We can achieve a reduction in energy demand either by improving the energy efficiency of buildings, lighting and appliances, or by changing the way we behave so that we use energy more intelligently and reduce the amount we need.*' [6].

Past decade has seen an increase in the evaluation of energy use in buildings. This has brought to notice that there is a considerable gap between the predicted and actual energy consumption in the investigated buildings. The authors of this paper have conducted a study analysing the thermal images of residential buildings on two winter night, when temperature was between 3–5 °C. It was found that no matter how insulated the building is, occupants' behaviour of opening windows results in considerable loss of heat [7]. There has been an increasing evaluation of energy use in buildings in the past 15 years, and it has been widely acclaimed that there is a considerable gap between the predicted and actual energy consumption in buildings. Occupant behaviour is one of the most overlooked parameter during energy efficiency design of buildings [8, 9].

In a project [10] aimed at reducing domestic energy usage by 20%, by exploring the relationship between the fabric of houses, heating systems and occupants' behaviour that work towards optimum comfort levels and the energy usage in the process, it has been found it is a complex problem and that there are significant variabilities between and within households over short and lengthy periods. They have found that thermostat settings vary from as low as 15 °C to as high as 30 °C. Although energy costs have been cited as a source of concern, the setting is found to be based on comfort rather than cost for most participants. They also have found that central heating is used in many different ways; some of them adjusted the thermostat directly, some set timers and some turned the whole heating system on and off as required [10]. Bălan et al. used a rule-based control, in the simulations of the thermal model of a house, where the influence of occupancy is determined as a secondary heat input, therefore impacting the internal load [3].

Opening and closing of windows as an indication status of occupants' behaviour in a building could be very critical in energy consumption patterns as the authors have found that two houses with the same degree and type of insulation perform significantly different due to the difference in the occupants' behaviour of window opening and window closing [7]. This implies that window opening behaviour has significant influence on the energy consumption and it could be as important as thermal insulation. In this paper, a survey is presented to explore the response to window's opening in order to understand further the behaviour of people to support a better building design.

29.2 Methodology

To explore the impact of occupants' behaviour on energy efficiency of a building, with the focus on opening of windows, an on-line survey was conducted. It examined the windows-opening behaviour of people and its consequent effect on the energy efficiency of buildings. The survey aimed at people residing in the UK. A survey link, containing a brief information on the aim of the project and the questions, were sent through emails and social media groups to a collective of people across the UK. An ethical approval process was followed during this research work. For the quantitative measures, the survey was designed with a confidence level of 95% following a sample size of 195 respondents, with an error margin of 7%.

29.3 Survey Details

Region: The response included residents from all parts of the UK, of which 56.7% were from the east midlands and the rest were scattered across the UK (9.8% from greater London, 98% from South East England, 8.2% from West Midlands, etc.)

Type of house: The response included all types of houses; 33.8% live in semi-detached houses, 25.6% in detached houses, 24.1% in flats and 12.8% of the respondents lived in terraced houses. While 52.8% of the people live in their own houses, 90% of the people who did the survey pay their own electricity and gas bills.

Ethnicity: 51.8% of the respondents were from Asian British background, while 36.5% were from white British/other white background; and the rest were from other ethnicities (Black, Arab, other multiple ethnic groups.)

Age range: The survey was answered by people from all age groups, as shown in Table 29.1.

Statistical analysis is carried out to analyse the patterns and the frequency of people opening their home windows. It is clear that if occupants frequently open or

Table 29.1 Percentage of respondents for different age groups

Percentage of respondents (%)	Age group
9.7	18–25
31.8	26–34
30.3	35–44
17.4	45–54
8.7	55–64
2.1	65–74

keep windows open when the heating system is on, this will have a negative effect on the energy conservation due to air infiltration.

29.4 Results and Discussion

The survey was distributed from November 2019 to February 2020. A total of 195 responses were received. Figure 29.1 shows the frequency of opening windows, by the respondents. It can be seen that although the frequency of window opening varied, 90.1% of the respondents open their house windows and only 9.9% never opened their house windows. Of the people who opened windows, 45.6% open windows at least once a day. About 44% of the people open windows for at least 30 min a day, of which 16% leave it open for 2–5 h. About 77.8% of people open windows when the heating is ON, of which 17% open it very frequently (see Fig. 29.2).

The most common reason to open windows in any room in a house is to get some fresh air. Apart from that, people open kitchen windows to get rid of cooking odour; and they open bathroom windows due to condensation.

A relevant point to note is that although 86.5% of the people who have done the survey are graduates or post-graduates, 45.5% of the people do not know the type of insulation in their homes, which is evident from Fig. 29.3. This shows that the behaviour of the people is independent of their knowledge in relation to their building's insulation.

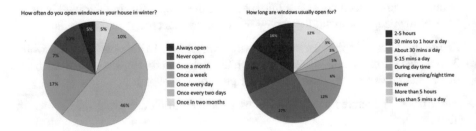

Fig. 29.1 Frequency of opening windows

How often do you open windows when the heating is ON?

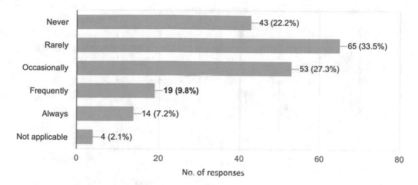

Fig. 29.2 Frequency of opening windows when heating is ON

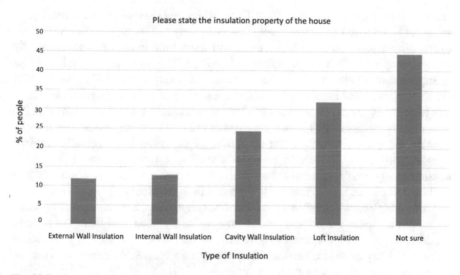

Fig. 29.3 Type of insulation as known to occupants

From the survey, it is evident that opening of windows is mainly a personal behaviour. For instance, let us analyse the response of participant 119. He/she prefers have the ambient temperature of the house to be between 18–22 °C. It is a detached house, with external wall insulation, internal wall insulation, cavity wall and loft insulation. In spite of this, the occupant opens the windows every day and for about 30 min. On the other hand, participant 77 who prefers the same ambient temperature of 18–22 °C, lives in a new built with cavity wall insulation and loft insulation; but leaves the windows closed at all time.

It can be seen that 77.8% of people open windows when heating is ON, 17% of which open windows very frequently. The most common reason for opening

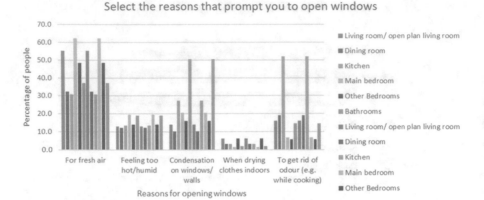

Fig. 29.4 Reasons for opening windows

windows is for fresh air (see Fig. 29.4). Other important reasons are condensation on walls and to remove odour while cooking. But energy efficiency in buildings have several dimensions. Children spend more and more time in their rooms, especially during the current pandemic situation of Covid19. The air quality therefore plays an important role in the quality of life and must be taken seriously [2]. The RSPCH and Royal College of Physicians conducted a study and provided a report based on the systemic review of the science of indoor air pollution. It has been found that the current energy efficiency of housing measures are reducing ventilation and thereby reducing the quality of air, putting the health of the residents, especially children at stake, particularly those with conditions such as asthma. Although residential building design and construction has evolved to emphasise energy efficiency and insulation, it can be argued that ventilation has been overlooked [11].

There is not enough consistent data on indoor air across the national housing stock [2]. Indoor air quality and energy efficiency are inter-related and there is a need to be addressed concurrently, for it to have any useful impact. Also, energy efficiency is one of the key aspects related to fuel poverty, as a household's fuel costs can be reduced based on increasing its energy efficiency [12].

29.5   Conclusion

Energy efficiency of buildings is influenced by several factors such as quality of the building envelope, wall insulation and occupant's behaviour. Occupants' behaviour is an overlooked aspect in many cases. The conducted survey shows that:

- Occupants' behaviour could have considerable effect on energy savings in a house, due to opening of windows which will cause air infiltration and reduce the effect of envelope's insulation.
- Further studies are still needed to analyse how the issue of windows opening can be addressed and the effect of that on energy consumption and health in terms of indoor air quality.

One way to address thermal insulation, while maintaining fresh indoor air quality, is by considering heat-recovery ventilation systems as a key aspect when planning new buildings or retrofitting of old ones. Also, more research is needed in relation to air quality and its link to indoor and outdoor interface and the influence on health.

References

1. Climate change: UK government to commit to 2050 target—BBC News (2019). https://www.bbc.co.uk/news/science-environment-48596775. Accessed 18 Oct 2019
2. S. Holgate et al., The inside story: health effects of indoor air quality on children and young people, pp. 1–96 (2020)
3. B.F. Balvedi, E. Ghisi, R. Lamberts, A review of occupant behaviour in residential buildings. Energy Build. **174**, 495–505 (2018)
4. International Energy Agency, Total Energy Use in Buildings: Analysis and Evaluation Methods (Annex 53) (2016)
5. Innovate UK, Building Performance Evaluation Programme: Findings from domestic projects, Swindon (2016)
6. Department of Energy and Climate Change, The Carbon Plan: Delivering our low carbon future (2011)
7. S. Salim, A. Al-Habaibeh, The Effect of Insulation on energy savings in residential buildings—myth or reality?, in *The International Conference on Energy and Sustainable Futures (ICESF)* (2019)
8. K. Schakib-Ekbatan, F.Z. Çakıcı, M. Schweiker, A. Wagner, Does the occupant behavior match the energy concept of the building?—analysis of a German naturally ventilated office building. Build. Environ. **84**, 142–150 (2015)
9. G.T. Wilson, T. Bhamra, D. Lilley, Reducing domestic energy consumption: a user-centred design approach, in *Knowledge Collaboration & Learning for Sustainable Innovation ERSCP-EMSU Conference, Delft, The Netherlands, 25th–29th October 2010,* 2010, pp. 200–222
10. D. Shipworth, Carbon, Control and Comfort I UCL Energy Institute—UCL—London's Global University (2012)
11. R. Morrison, Energy-efficient housing is reducing ventilation and trapping air pollution indoors, *Mailonline*
12. E. and I. S. Department of Business, Annual Fuel Poverty Statistics in England, 2020(2018 Data) (2019)

Chapter 30
Fuel Poverty and Health Implications of Elderly People Living in the UK

Ali Mohamed Abdi, Andrew Arewa, and Mark Tyrer

Abstract Fuel poverty is widely recognised as distinct form of injustice and social inequality and a front burner issue in the last three decades in the UK. The crisis affects 4.5 million households in the UK, and it is a major high-risk contributor to health of elderly people (NEA in Effects of Living in Fuel Poverty, NEA.ORG, London, 2020, [1]). Thus, the consequences of fuel poverty range from psychological stress, worry and isolation to serious health conditions such as respiratory and circulatory diseases. The aim of the study is to investigate the role of fuel poverty on reoccurring health risks of elderly people. The study adopted quantitative research methods with participants drawn from West Midlands region of England - UK; an area with high population of elderly people, carers, health professionals and energy professionals. Findings from the survey indicate that fuel poverty is one of the major aspects that contributes to health implications among elderly people in the UK.

Keywords Fuel poverty · Health risks · Excess winter deaths · Energy efficiency

30.1 Introduction

Fundamentally, a household is in fuel poverty when occupants face high cost of keeping adequately warm due to low income. Hence, the three main issues that determine fuel poverty are household income, cost of energy and energy efficiency. Numerous academic literatures on fuel poverty mainly focus on policy aspect and affordable/cold homes links to health problems [2]. However, research regarding the role of fuel poverty on health risks of elderly people is scarce. The most recent data by National Energy Action (NEA) [1, 3] reported that there were 4.5 million families in the UK living in fuel poverty. The aim of this paper is to investigate whether fuel poverty is one of the factors that contributes to health issues among the elderly people.

A. M. Abdi (✉) · A. Arewa · M. Tyrer
Faculty of Engineering Environment & Computing, Coventry University,
Priory Street, Coventry CV15FB, UK
e-mail: abdia7@uni.coventry.ac.uk

© The Author(s) 2021   241
I. Mporas et al. (eds.), *Energy and Sustainable Futures*, Springer Proceedings in Energy,
https://doi.org/10.1007/978-3-030-63916-7_30

30.2   Fuel Poverty and Health

Fuel poverty is having to spend 10% or more in household net income to heat their homes to an adequate standard of warmth [2]. Subsequently, there is fine line between fuel poverty and living in poverty. The former links to shortage of income or wealth whilst the latter happens when household's resources are well below their minimum needs [1].

Relationship between fuel poor and health has been subject of much debate in the UK for many years with most research linking the problem to cold dump houses usually occupied by elderly people. However, most research does contribute to knowledge about elderly people cases and what happens to their health when they live in cold or less energy efficient homes [4, 5]. The negative impact of cold temperature is mostly felt by older people and it is often exacerbated by pre-existing health conditions [3]. Consequently, elderly people are at higher risk of increased blood pressure and blood coagulation both linking to low temperature and can lead to respiratory and circulatory conditions [6]. In winter period, elderly people who are between the age bracket of 65 and over are subject to the greatest increase of deaths. For example, in 2017/18 winter period there were 50,100 excess winter deaths (EWDs) in the UK [7]. Table 1 below shows excess winter deaths from 2015 to 2018.

30.3   Research Methods

The results from this paper is drawn on findings from quantitative survey of elderly people, health professionals and energy professionals living in West Midlands region in the UK. The main reason for choosing the area is due to its high population of

Table 1 Excess winter deaths on Elderly People (ONS 2019)

		Excess winter deaths		
Sex	Age	2015/2016	2016/2017	2017/2018
Male	0–64	1,730	1,350	2,700
	65–74	1,820	1,790	3,200
	75–84	3,370	4,440	6,700
	85+	4,290	6,810	9,300
	All ages	11,200	14,390	21,900
Female	0–64	1,200	890	1,500
	65–74	1,520	1,850	2,500
	75–84	3,520	4,690	6,900
	85+	7,140	12,720	17,300
	All ages	13,380	20,140	28,100
	Total	24,580	34,530	28,100

low-income people and high number of excess winter deaths compared to other parts in the UK [7]. Questionnaire administered included both online and face to face with each category measuring the study objectives. Stratified random sampling was used to select the study quantitative data.

Apart from quantitative approaches, archive data about Excess winter deaths and temperature from 2014-2019 were obtained and analysed in order to have a broad view in how low temperature during winter affects elderly people health. The archive data were retrieved from Office of National Statistics (ONS) and UK's Meteorological office website. Months used for the winter temperature are December, January, February and March.

30.3.1 Quantitative Data Analysis

Total of 100 questionnaires were administered, 92 were returned and 8 were considered invalid because they were completed incorrectly. The 92 returned questionnaires focussed in three areas; elderly people accommodation and energy affordability, energy professional tackling of fuel poverty and health professionals reporting on common health problems associated with elderly people. In line with their status and discipline different types of questions were asked to the participants. For instance, with the elderly people, the questionnaire asked whether they prioritise energy bills before food and clothing. 41% and 34% of the respondents ticked strongly agree/agree that they prioritise their energy bills before buying essentials such as food and clothing. Figure 30.1 shows the prioritising of energy bills before other basic needs by the elderly people.

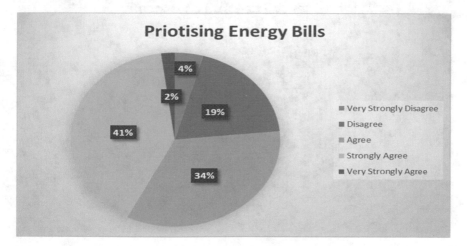

Fig. 30.1 Prioritising energy bills over other basic needs

The health professionals were asked about the most common reported cold related disease among elderly people. Respiratory and influenza was ticked as the most common cold related disease among elderly people. Figure 30.2 illustrates the most common cold related diseases among elderly people.

Lastly, energy professionals were asked whether high energy price and low income contribute to fuel poverty. 43% and 39% of the respondents strongly agree/agree that

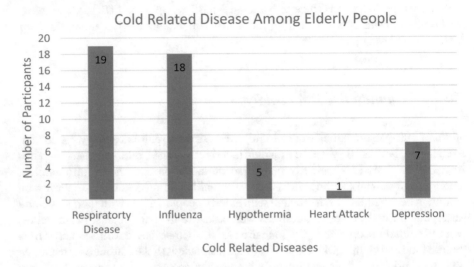

Fig. 30.2 Most common cold related diseases in elderly people

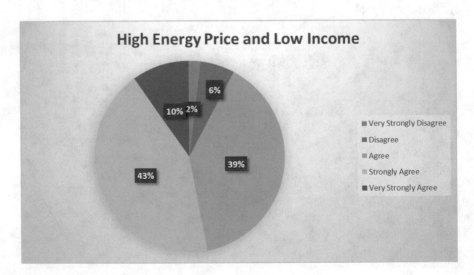

Fig. 30.3 High energy price and low-income contribution to fuel poverty

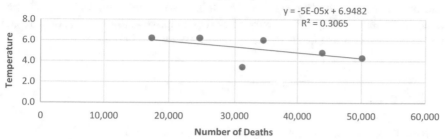

Fig. 30.4 Excess winters deaths and winter monthly temperature (ONS 2019)

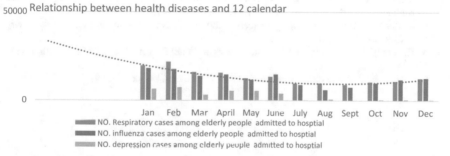

Fig. 30.5 Relationship between health disease and 12 calendar months

high energy price and low income contribute to fuel poverty. Figure 30.3 shows contribution of fuel poverty by issues such as high energy price and low income.

Additionally, correlation analysis was conducted using ONS archive data for years 2014 to 2019 excess winter deaths and winter month temperature between the period. The correlation coefficient between EWDs and winter month temperature produced a value of 0.63; meaning there is a relationship between the winter months temperature and EWDs. Figure 30.4 shows excess winter deaths and winter monthly temperatures for the years 2014 to 2019. Meanwhile, Fig. 30.5 illustrates relationship between health disease and 12 calendar months.

30.4 Conclusion

The paper investigated the role of fuel poverty in health risk of elderly people in the UK. Evidence from the survey reveal that study participants have perceived view on strongly agree that fuel poverty plays significant role in health risks of most elderly people. Besides, ONS archive used for correlation analysis of diseases, excessive death and climatic conditions show that there is relationship between cold condition and health risk of elderly people. Hence, this echoes the prompt development of housing strategy with specific energy efficiency adaptive features, cost effective and universally available to meets the needs of elderly people in the UK.

References

1. NEA, Effects of Living in Fuel Poverty, NEA.ORG, London, 2020
2. R. Moore, Definitions offuelpoverty: implicationsforpolicy. Energy Policy **40**, 19–26 (2012)
3. J. Hills, Getting the measure of fuel poverty: final report of the Fuel Poverty Review, London School of Economics and Political Science, London (2012)
4. A. Kearnes, E. Whitley, A. Curl, Occupant behaviour as a fourth driver of fuel poverty (aka warmth & energy deprivation). Energy Policy **129**, 1143–1155 (2019)
5. C.W. Price, K. Brazier, W. Wang, Objective and subjective measures of fuel poverty. Energy Policy **49**, 33–39 (2012)
6. M.R. Team, The Health Impacts of Cold Homes and Fuel Poverty, Friends of the Earth, London (2011)
7. ONS, Excess Winter Mortality in England and Wales, ons, London (2019)

Chapter 31
How Clean Is the Air You Breathe? Air Quality During Commuting Using Various Transport Modes in Nottingham

Bubaker Shakmak, Matthew Watkins, and Amin Al-Habaibeh

Abstract Air quality has developed into a significant global issue and its negative effect on human health, wellbeing and ultimately the effect of shortening of life expectancy is becoming a pressing concern. Such concerns are most acute in cities in the UK. Although many cities, including Nottingham, are taking significant measures to enhance air quality, there was limited work focusing on the individual's experience during commuting. This paper suggests a novel approach for measuring commuting air quality through quantifying particulate matters PM2.5 and PM10, using the city of Nottingham as a case study. Portable low-cost systems comprising of a GPS sensor and an Aeroqual pollution data logger were used to capture data and develop the sensor fusion via newly developed software. Data was collected from a variety of transport modes comprising bike, bus, car, tram and walking to provide evidence on relative particulate levels and 2D and 3D data maps were produced to communicate the relative pollution levels in a publicly accessible manner. The study found as expected particulate pollution to be higher during peak hours and typically closer to the city. However whilst the lowest particulate concentrations were found on the Tram the highest were for cyclists contrary to the literature. The project encompasses a democratic crowd sourced approach to data collection by enabling the public to gather data via their daily commute, increasing people's awareness of the air quality in their locality. The acquired data permitted a range of comparisons considering differing times of day and zones such as the city centre and surrounding residential areas in the City council boundary.

Keywords Air quality · Monitoring · Transport mode · Mapping · Crowd sourcing

B. Shakmak · M. Watkins (✉) · A. Al-Habaibeh
School of Architecture and the Built Environment, Nottingham Trent University, Nottingham, UK
e-mail: Matthew.Watkins@ntu.ac.uk

B. Shakmak
e-mail: Bubakar.Shakmak@ntu.ac.uk

A. Al-Habaibeh
e-mail: Amin.Al-Habaibeh@ntu.ac.uk

© The Author(s) 2021
I. Mporas et al. (eds.), *Energy and Sustainable Futures*, Springer Proceedings in Energy,
https://doi.org/10.1007/978-3-030-63916-7_31

31.1 Introduction

The dangers of air pollution are well documented and exposure to the particulate matter in the air is a significant contributor to cardiovascular disease and can lead to an increase in premature death rates [1, 6]. Air quality monitoring is therefore important in regard to protecting human health but also the environment [2].

Nottingham is one of 44 UK cities that exceeded safe levels of air quality in 2017 [3, 6]. The World Health Organisation (WHO) states that particles (PM 2.5) in the air should not exceed 10 micrograms per cubic metre, whilst in Nottingham, it was 12 micrograms per cubic metre at its peak [3]. Over recent years Nottingham has significantly reduced its air pollution levels by investing in public transport, active travel infrastructure and discouraging private car use through a workplace parking levy [4, 6]. However, in order to continue to drive down air pollution public engagement is also important. Public engagement could be increased through awareness of the issues and this project proposes a potential solution by utilising public crowdsourced data to continually update a pollution map to identify the most polluted areas, for use as an early alarms system and to identify and address pollution hotspots. This project aims to produce and test an inexpensive, easy to use, yet effective technology to monitor pollution and air quality around the city of Nottingham, utilizing citizens to collect data through their daily commute and leisure travel. Particulate Matter (PM) data collected with GPS coordinates will permit the creation of pollution maps, by location and transport mode.

31.2 Methodology

This study utilised modified backpacks shown in Fig. 31.1 each containing an Aeroqual Series 500 air quality monitor with a particulate matter sensor able to measure PM 2.5 & PM 10, with an ITrail GPS coordinate tracking unit enabling the data to be analysed and overlaid in processing later. The choice of both the PM sensor and

Fig. 31.1 Depicting the integrated device placed inside a backpack

GPS unit were made according to accuracy, low weight, and portability and battery life. Volunteers were asked to carry the backpack with them for 1–2 days each and volunteers were selected to replicate journeys from different areas into the city centre using a variety of transport modes including walking, cycling, bus, car and tram. The data was collected in three months between April and July 2019. Between them the volunteers covered approximately half of the Nottingham city area.

31.3 Pollution Data Processing

The inclusion of the GPS tracker permitted the location coordinates to be recorded to an accuracy of ±3 m and aligned and processed with the pollution data by matching the time and date stamps from each device within Matlab software, at which point the mode of transport would be added from notes.

Proprietary software was created which permits the classification of the GPS coordinates to align with six pre-determined city zones. The six zones were determined based on the location of the participants commute and the distance from the city centre, growing between 1 & 2 miles in diameter as shown in Fig. 31.2.

The data was thoroughly checked and cleaned to remove duplicates before being analysed and used to plot the pollution maps shown in Fig. 31.3. The Matlab code created enabled automatic sifting of the data to filter the data based on variables and associated categories to carry out practical comparisons. The data has been marked for each transportation method in order to carry out some sort of comparison to identify the safest mean of transportation for a human without being exposed to extra-polluted air.

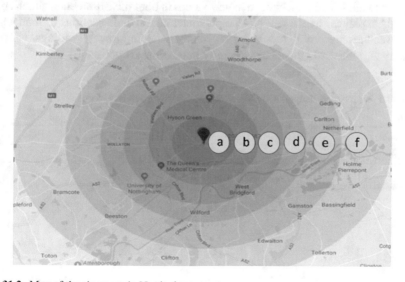

Fig. 31.2 Map of the six zones in Nottingham

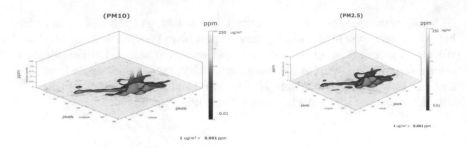

Fig. 31.3 Particulate (PM10 & PM2.5) Relief Maps of Nottingham

31.4 Results and Discussions

The processed data was converted and overlaid on maps to create a 3D relief for both PM10 and PM2.5 pollution types, visually representing the status of pollution around the city. These maps illustrate the paths used by the volunteers to collect the samples of the pollution data and the amount of pollution around the city for PM10 and PM 2.5 as shown in Fig. 31.3 respectively.

The range of samples recorded by transport mode as shown below in Table 31.1, whilst the sample was especially high for walking this is to be expected as with all transport modes an element of walking is undertaken at either end of the trip and even between modes in some cases. Furthermore due to this being a city centre location walking as the main mode is more likely, especially amongst student participants.

The travel modes were compared and an average taken across each as shown in Fig. 31.4. The Bus had the highest PM 10 values, whilst cycling the highest PM2.5 values. This is surprising considering that by this time much of the council bus fleet is Biogas powered. The tram had the lowest values in both measurements, which could be partly explained by the fact that Trams routes can include sections of road where motor vehicles are prohibited. However cars had the 2nd lowest readings, which is contrary to published guidance, especially in relation to cyclists [5]

Table 31.1 Number of samples and the percentage taken by each transportation mode

Mode	Walk	Bicycle	Car	Bus	Tram
No samples	3108	291	243	359	195
Percentage (%)	74	7	6	9	5

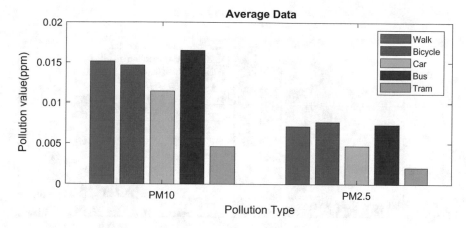

Fig. 31.4 The mean PM data acquired by transportation modes

Table 31.2 Time frame categories

From	To	Description	Time Interval
00:00	06:00	Quiet morning time	A
06:00	08:00	Early morning hours	B
08:00	10:00	Rush hours	C
10:00	16:00	Working hours	D
16:00	18:00	Evening rush hours	E
18:00	21:00	Evening social hours	F

31.4.1 Effect of Time of Day

To gain further insight Matlab analysis considered the timestamps dividing the data into seven categories as shown in Table 31.2, to consider the effects of road congestion. As expected Fig. 31.5 shows pollution is typically higher during rush hours periods, but this appears to continue even later than expected until 9 pm.

31.4.2 Zonal Considerations

By comparing the pollution across the 6 zones (introduced in Fig. 31.2), Fig. 31.6 demonstrates that the amount of pollution in the centre of the city is high with the pollution reducing towards the outskirts of the city. The average of both PM10 and PM2.5 shows the pollution values are higher in the city centre and typically decreasing with distance outside of the city. However Zone e appears to be an

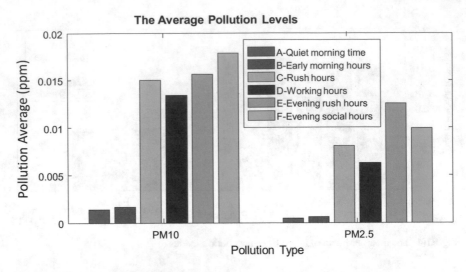

Fig. 31.5 The average pollution data by time of day

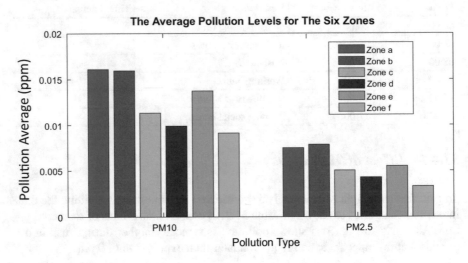

Fig. 31.6 The average of the pollution in the six zones around Nottingham

anomaly, but upon investigation appeared there was a higher deviation in the readings from Zone e which could account for this and so a larger and longer term sample needs to be conducted to explore this further.

31.5   Conclusions

The study suggests that a crowdsourced approach to monitoring air quality levels is viable and has potential to be highly effective. The results typically follow expected patterns with higher levels typically closer to the city centre and higher concentrations at peak times. However the results also highlight some surprising results in relation to the transport mode and further research considering this would be beneficial. Of particular interest is the highest level of PM2.5 when cycling, which combined with the additional aerobic exercise, requiring deeper breathing could have pose long-term health issues with particulates entering the lungs.

31.6   Further Work

The team will continue this research, focussing specifically in relation to cycling exploring air quality over longer durations, with a number of weeks per participant repeating a set commute to and from set key locations into the city centre. This work will consider a wider range of air pollutants using multiple sensors and will be repeated at different times of the year to consider the effects of seasonal changes. It is also hoped to develop a dedicated application to be used in mobile phones when connected to the device to manage and send the data remotely to the network.

Acknowledgements This study was funded via the NTU Sustainable Futures Seedcorn.

References

1. R.D. Brook et al., Particulate matter air pollution and cardiovascular disease: an update to the scientific statement from the american heart association. Circulation **121**(21), 2331–2378 (2010)
2. Why Air Quality is important, *Oxfordshire AirQuality*. [Online]. https://oxfordshire.air-quality.info/why-air-quality-is-important. Accessed 18 Sep 2019
3. J. Pritchard, "Nottingham air does not meet safe levels says major new report, *Nottingham Post* (2017). https://www.nottinghampost.com/news/nottingham-news/nottingham-air-too-dangerous-breathe-704643. Accessed 29 June 2020
4. C. Reid, Nottingham's Workplace Parking Levy Creates Jobs, Cuts Car Use and Slashes Pollution, *Forbes* (2019). https://www.forbes.com/sites/carltonreid/2019/10/17/nottinghams-workplace-parking-levy-creates-jobs-cuts-car-use-and-slashes-pollution/. Accessed 24 June 2020
5. A. Karanasiou, M. Viana, X. Querol, T. Moreno, F. de Leeuw, Assessment of personal exposure to particulate air pollution during commuting in European cities—Recommendations and policy implications. Sci. Total Environ. **490**, 785–797 (2014)
6. Nottingham City Council, Nottinghamshire Air Quality Strategy 2019–2028 (2019). https://committee.nottinghamcity.gov.uk/documents/s97117/Enc. 1 for Draft revised Nottingham and Nottinghamshire Air Quality Strategy.pdf. Accessed 29 June 2020

Part VI
Materials

Chapter 32
Reformulating Ceramic Body Composition to Improve Energy Efficiency in Brick Manufacture

G. Wie-Addo, A. H. Jones, S. Palmer, V. Starinieri, J. Renshaw, and P. A. Bingham

Abstract The influence of inorganic minerals (colemanite and nepheline syenite) as additives for sustainable clay brick manufacture has been examined. Each additive was added at 4 wt% to 96 wt% brick clay and samples were fired to 950 °C and 1040 °C and then compared with samples of 100% brick clay. Multiple analytical techniques (X-ray fluorescence, dilatometry, boiling water absorption, volumetric shrinkage, and mercury porosimetry) were used for analysis. Dilatometry shows that the additives influenced the temperature at which shrinkage began and the extent of that shrinkage. The use of colemanite reduced the temperature at which the shrinkage began by 120 °C and nepheline syenite reduced it by 20 °C. A linear shrinkage in dilatometry of 1% (from the maximum expanded length) was achieved at 1000 °C for 100% clay, 875 °C for colemanite additions and 970 °C for nepheline syenite additions. However, for samples fired at 1040 °C for 2 h colemanite containing samples had significantly lower volumetric shrinkage and higher water absorption than 100% clay and nepheline syenite samples, suggesting the presence of higher amounts of open porosity caused by the decomposition of the colemanite on heating. Samples containing nepheline syenite had a lower volumetric shrinkage but also a marginally lower water absorption than the 100% clay. The further optimisation of these or similar additives could potentially provide energy saving opportunities and reductions in CO_2 emissions for brick manufacturers.

Keywords Clay bricks · Energy savings · Inorganic mineral additives

G. Wie-Addo (✉) · A. H. Jones · V. Starinieri · P. A. Bingham
Materials and Engineering Research Institute, Sheffield Hallam University, Sheffield S1 1WB, UK
e-mail: gloria.wie-addo@student.shu.ac.uk

V. Starinieri
e-mail: Starvinc@gmail.com

S. Palmer · J. Renshaw
Wienerberger Ltd, Wienerberger House, Brooks Drive, Cheadle Royal Business Park, Cheshire SK8 3SA, UK

© The Author(s) 2021   257
I. Mporas et al. (eds.), *Energy and Sustainable Futures*, Springer Proceedings in Energy,
https://doi.org/10.1007/978-3-030-63916-7_32

32.1 Introduction

Energy is an important factor for the ceramic sector, the key role of the thermal treatment of raw materials in activating the "physical-mineralogical transformation processes" to create the desired properties [1].

Gas combustion, which accounts for 70–80% energy use in ceramic manufacture contributes substantially to the overall CO_2 and greenhouse gas (GHS) released into the atmosphere. These are major contributors to global warming and if left unabated might result in devastating harm to the environment. This has brought about the commitment to minimize CO_2 emissions in the ceramic sector through the 2050 decarbonisation plan which intends to address three major categories significant to the ceramic sector. Thus, the raw materials, energy efficiency and CO_2 reduction are addressed here through breakthrough technologies of which raw materials and processing are key [2].

This paper focuses on making changes to the raw materials used to make bricks (the clay body) to meet the CO_2 emission reduction targets set by the de-carbonation plan. Varying the clay composition using additives is an established way to alter clay properties such as plasticity, colour etc. or, in this case, reducing energy consumption in manufacture. However, identifying the right additive while maintaining the minimum brick properties is not straight forward since international standard bodies such as ASTM and EU regulate these properties [3].

Among the various clay additives identified in the literature [4], the use of colemanite ($2CaO.3B_2O_3.5H_2O$) and nepheline syenite ($K_{1.37}Na_6(AlSiO_4)_8$) containing materials have gained much attention. This is due to their high fluxing content of fluxing elements including Ca, Na, Mg and K [5]. In clay body manufacture the presence of these fluxing agents can create lower melting point phases during firing, enhancing sintering behaviour and properties while using lower firing temperatures. The literature shows the effect of these additives' incorporation into semi-vitreous, glass and glass ceramic products [6–8]. The use of these additives enhanced vitrification, reduced internal stress caused by free quartz in clay and reduced the required firing temperature by 80 °C–120 °C. The use of 1% colemanite was able to reduce the firing temperature by 50 °C in porcelain bodies without affecting product quality [5, 7, 9–12]. However, the incorporation of the additives into clay brick composition has not been reported.

This study aims to assess the effect of adding 4 wt% of these inorganic additives on the firing behaviour and properties of clay bricks.

32.2 Experimental Procedures

Cretaceous age clay mined in the south-eastern part of UK, sourced from Wiener-berger Ltd and inorganic minerals, colemanite ($2CaO.3B_2O_3.5H_2O$) and nepheline syenite ($K_{1.37}Na_6(AlSiO_4)_8$) were separately dried in an oven at 120 °C for 48 h.

4 wt% of each dried additive was weighed and mixed with 96 wt% dried clay then milled in a Retzsch vibratory disc mill for 1 min at 800 rpm into a fine powder. The dry powders were mixed with 21% distilled water to provide sufficient plasticity for shaping. A hydraulic press with 7 tonnes force was applied to 8.3 g of the composition to form a 20 mm dimeter, 5 mm thick cylindrical pellet. After shaping, the green body dimensions were recorded in the "as-pressed" condition. Samples were then left on a bench and further dried in an oven (100 °C) for approximately 36 h to gradually eliminate water. The dry cylinders were fired to 950 °C or 1040 °C at 1 °C/min and held for 2 h at the top temperatures before cooling at the same rate.

Chemical analysis of the raw materials and the fired bodies using XRF fused bead technique was carried out as described elsewhere [13]. Dilatometry (NETZSCH DIL 402SE) was used to measure the linear dimensional changes between 35 °C–1050 °C with a heating rate of 5 °C/min. A rod of 25 mm length and 6 mm diameter of unfired material was used for dilatometry. Volumetric shrinkage measurement were determined on cylindrical (20 × 5 mm) test pieces in the wet, dry, and fired conditions. The temperature at which the shrinkage began was taken as the point at which the expansion rate (dL/dt) becomes negative. As a comparative measure of the shrinkage at the highest temperatures, the temperature at which each sample showed a 1% shrinkage from its maximum value was noted. Additionally, the level of shrinkage (measured from the maximum level) at 1000 °C was noted. The boiling water absorption of fired samples was measured according to BS EN 772-7. The boiling test results reported are the averages of 3 sample measurement. Mercury porosimetry was used to determine the intrudable pore volume and the distribution of pores sizes. The intrusion tests were carried out using Pascal 140/240 system with mercury angle at 140 degree.

32.3 Results and Discussions

32.3.1 Chemical Analysis

XRF showed that colemanite and nepheline syenite had low levels of Fe_2O_3 not greater than 1%. This result agrees with previous studies [12]. The XRF used for this analysis could not detect boron because the x-ray energy is below the instrument's minimum detection level. However, the level of boron present can be inferred from the formula for colemanite ($2CaO. 3B_2O_3. 5H_2O$) [14]. Nepheline syenite had a total of 18.9% fluxing oxides, including 8% Na_2O and 9.3% K_2O compared to a total of 3% in the clay without additives (excluding Fe oxides).

The chemical constituents of fired ceramics revealed that, 4 wt% colemanite (C4W) and nepheline syenite (N4W) have approximately the same SiO_2 content of 71–72 wt% as shown in Table 32.1. The additional percentage of fluxing oxides (Na_2O, MgO, CaO, K_2O) is expected to provide additional liquid phase during firing to improve densification at a lower temperature [15].

Table 32.1 X-ray fluorescence (XRF) analysis of raw materials and clays fired to 1040 °C with no additions; and with 4 wt% additions of colemanite (C4W) and nepheline syenite (N4W)

Raw Materials	SiO_2	Al_2O_3	Fe_2O_3	TiO_2	Na_2O	K_2O	MgO	CaO	SrO_2	SO_3	Others	SUM (%)	L.O.I
Raw material oxides/weight %													
Clay	73.5	15.4	6.1	1.5	0.0	2.2	0.5	0.3	–	0.2	<0.1	100	16.75
Colemanite	18.0	2.0	1.0	0.3	0.1	1.1	5.7	69.2	2.5	0.3	<0.1	100	29.50
Nepheline Syenite	55.7	24.3	0.1	0.2	8.0	9.3	0.0	1.6	–	0.3	<0.1	100	5.8
Fired ceramic oxides/ weight %													
Fired Clay	71.0	15.9	7.7	1.5	0.2	2.3	0.6	0.4	–	0.0	<0.1	100	4.83
C4W	71.8	14.9	6.9	1.5	0.2	2.1	0.6	1.5	–	0.0	<0.1	100	5.62
N4W	72.0	15.7	6.9	1.3	0.5	2.4	0.1	0.4	–	0.0	<0.1	100	6.49

32.3.2 Effects of Additives on Ceramic Properties

Dilatometric curves obtained with the push rod dilatometer are shown in Figs. 32.1a–c. The 100% clay sample has almost zero coefficient of expansion between the quartz inversion (581 °C) and around 840 °C and reached a maximum linear expansion (dL/L_0) of around 0.95%. This must mean that the sample is contracting due to sintering at about the same rate as it is expanding due to thermal expansion and suggests that a mechanism of shrinkage is active above 600 °C. The sample then shrinks rapidly, beginning at 820 °C and continuing up to the end of the test at 1050 °C. Shrinkage of 1% (from the maximum value of dL/L_0) was achieved at a temperature of 1000 °C.

For the clay with colemanite additions (Fig. 32.1b) the maximum expansion on heating was 1.5%, 50% higher than for the clay alone. The derivative curve shows two

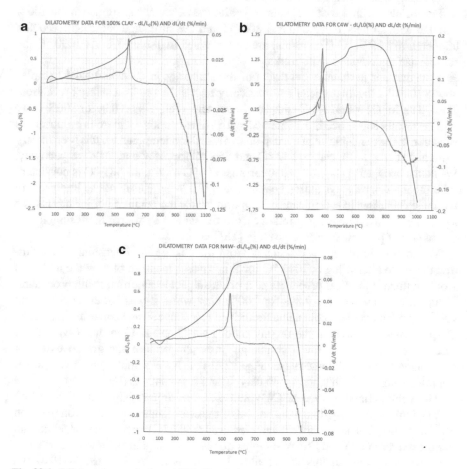

Fig. 32.1 Dilatometry results for **a** 100% Clay, **b** 4 wt% colemanite and **c** 4 wt% nepheline syenite, all heated to 1050 °C at 5 °C/min

Table 32.2 Data obtained from dilatometry experiments showing the effect of additives on the temperature for the onset of shrinkage and extent of shrinkage

Sample	Temperature for onset of shrinkage (°C)	Temperature for 1% shrinkage (°C)	Linear shrinkage at 1000 °C (%)
100% Clay	820	1000	1
C4W	700	875	3
N4W	800	972	1.4

sharp peaks at 363 °C and 380 °C which are due to the decomposition of colemanite by the removal of the inter-crystallite water as reported by [10–16] leaving a glassy borate structure [14] and resulting in a rapid expansion of the body. Shrinkage begins at 700 °C and a shrinkage of 1% was achieved at a temperature of 875 °C. At 1000 °C the sample had undergone 3% shrinkage.

For the clay with 4 wt% Nepheline syenite the behaviour is similar to the 100% clay but with shrinkage beginning at 800 °C and 1% shrinkage being achieved at a temperature of 972 °C. The results are summarised in Table 33.2. At 1000 °C the sample had undergone 1.4% shrinkage.

Note that due to the rate and extent of the shrinkage in C4W and N4W samples above 1000 °C the dilatometer automatically aborts the run as a safety precaution. Thus, data for C4W and N4W ends at a lower temperature than that for 100% clay.

The results above need to be treated with some caution as there is an observed variation in the temperature of the α-β quartz transition temperature. It is seen to vary from 581 °C for 100% clay to 542 °C and 555 °C for N4W and C4W, respectively. While the accepted value for pure quartz is 573 °C a value of 581 °C is consistent with other work on clays at this heating rate [17]. What is difficult to interpret are the large decreases in α-β transition temperature for the ceramic bodies fired at the same rate when the quartz is primarily from the same clay source in all samples. A discussion of possible causes is given in [18].

Volumetric shrinkage, as calculated from the change in dimensions of a cylindrical sample after firing at 1040 °C for 2 h, gives a measure of how the additives would influence the final properties of a brick subject to a normal firing procedure. Volumetric shrinkage was 9.8% for 100% clay while it was lower, at 8.1% and 4.9% for nepheline and colemanite additives respectively, as shown in Fig. 32.2a. Despite small increases in the fluxing oxides Na_2O and K_2O in the N4W sample (see Table 32.1) and the lower shrinkage onset temperature and greater extent of shrinkage shown in dilatometry (see Fig. 32.1b) the N4W samples underwent lower volumetric shrinkage than the 100% clay after 2 h at 1040 °C (Fig. 32.1c).

The volumetric shrinkage of C4W samples was around half that of the 100% clay and this is probably due to the effects of the colemanite decomposition between 300 °C to 400 °C. The data for the samples fired at 950 °C suggest that the additional fluxing oxides in C4W and N4W samples do result in a higher volumetric shrinkage than the 100% clay at this temperature but this is not maintained to 1040 °C.

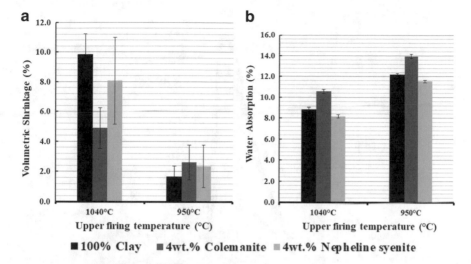

Fig. 32.2 **a** Volumeteric shrinkage and **b** boiling test result of 100% clay and clay with 4 wt% inorganic additives fired to 950 °C and 1040 °C

At 950 °C the volumetric shrinkages for C4W and N4W are significantly lower than those typically observed for bricks with the required properties as shown in Fig. 32.2a.

For water absorption of samples fired at 1040 °C, N4W recorded the lowest absorption of 8.2% compared with 100% clay absorption of 8.8% and C4W had highest absorption of 10.6% (Fig. 32.2b). These are mostly consistent with the volumetric shrinkage data where lower shrinkage suggests a higher volume of pores in the structure which are able to absorb water. At the lower temperature of 950 °C the water absorption (Fig. 32.2b) is higher for all samples, again consistent with the lower volumetric shrinkage results and expected higher porosity at 950 °C (Fig. 32.2a).

It was noted that the dilatometry results and the volumetric expansion results appear to suggest contradictory behaviour. From dilatometry the addition of colemanite and nepheline syenite both appear to reduce the temperature at which shrinkage starts and the temperature at which 1% linear shrinkage is achieved. However, these differences do not seem to have influenced the volumetric shrinkage of the samples after firing at 1040 °C for 2 h. For C4W samples the decomposition of the chemically bonded water in colemanite has resulted in a large expansion of the sample prior to the shrinkage and this has resulted in overall lower volumetric shrinkage. Other studies have reported a bloating effect at higher temperature as a major effect from the decomposition of calcite in colemanite which results in the release of CO_2 gas [19]. For N4W samples, the additional fluxing oxides appear not to have increased the volumetric shrinkage at 1040 °C and in fact the shrinkage is lower than the 100% clay.

Results for the porosity size distribution and porosity volumes as measured by mercury porosimetry on samples fired at 1040 °C and 950 °C are shown in Figs. 32.3 and Table 32.3. For 1040 °C firing C4W records the lowest total intrudable porosity

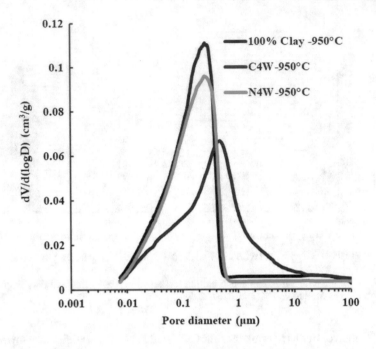

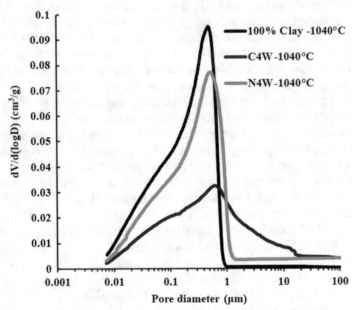

Fig. 32.3 Pore size distribution of fired clay with 4 wt% inorganic additives compared to 100% clay fired to 950 °C and 1040 °C

Table 32.3 Total intrudable porosity of fired clays with 4 wt% inorganic mineral additions compared to fired 100% clay

% Porosity by mercury intrusion porosimetry (MIP)		
Sample ID	Fired 950 °C	Fired 1040 °C
100% Clay	18.86	14.20
C4W	18.44	12.61
N4W	17.58	14.06

of 12.61% compared to 14.2% for 100% clay and 14.06% for N4W. This is in contradiction with the volumetric shrinkage and water absorption results where C4W has the lowest shrinkage and highest water absorption, suggesting that it should have around 20% more porosity than the 100% clay sample [20, 21].

In terms of pore size distribution (PSD) of samples fired at 1040 °C/2 h (Fig. 32.3), the 100% clay samples had all pores smaller than 1 μm. There is a bi-modal PSD present, with the main peak in pore size around 0.5 μm but a broad distribution of smaller pores sizes in the range 0.01–0.2 μm. N4W shows a very similar PSD but with some pores slightly larger than 1 μm and the peak slightly shifted to larger sizes. C4W has a very different PSD, with a smaller peak at 0.7 μm and a broad distribution between 0.01 μm and 11 μm. At the lower firing temperature of 950 °C the PSD of 100% clay and N4W are very similar to each other but are shifted to smaller pore sizes, from about 0.5 μm at 1040 °C down to 0.25 μm at 950 °C. The C4W PSD at 950 °C is similar to that at 1040 °C but with a shift in the main peak to smaller pore sizes (0.7 μm at 1040 °C and 0.5 μm at 950 °C) and there appear to be fewer large pores (in the range 8 to 10 μm).

Nicholson and Ross [22] explain that in clays pore radius can increase with firing temperature while the volume of porosity decreases with a change from 0.55 μm and 1.2 μm for buff fired clays fired to 900 °C and 1000 °C respectively.

32.4 Conclusions

The use of colemanite or nepheline syenite as an addition to brick making clay has been studied and the influence of these additives on the thermal and physical properties of the resulting fired clay bodies has been investigated. The following conclusions can be drawn:

1. The use of 4 wt% colemanite mineral additives to brick clays resulted in the production of decomposition products (OH) in the temperature range 300 °C–400 °C which created a large and rapid expansion (0.75% linear over a 100 °C range) of the clay body. The introduction of Ca and B appeared to reduce the temperature for the onset of shrinkage by 120 °C and result in a linear shrinkage at 1000 °C three times higher than that of the 100% clay. However, when fired at 1040 °C/2 h the clay body containing colemanite produced only half the volumetric shrinkage

compared with the 100% clay (4.9% vs. 9.8%). It also had a higher average pore size and pores sizes between 1 μm and 10 μm, which were not observed in the 100% clay. As a result, the water absorption was higher than the 100% clay (11% vs. 9%).

2. The use of nepheline syenite at 4 wt% introduced small amounts of additional fluxing oxides (Na_2O, K_2O) which could be expected to enhance the sintering of the brick clay at the firing temperature through the formation of more lower melting point phases. Dilatometry showed a reduction of around 20 °C in the shrinkage onset temperature and 1% linear shrinkage was achieved 28 °C lower than for the 100% clay. The linear shrinkage at 1000 °C was 1.4%, which was higher than for the 100% clay. However, volumetric shrinkage of N4W fired at 1040 °C/2 h was lower than for the 100% clay (9.8% vs. 8.1%) but water absorption was also lower (8.2% vs. 8.8%). Pore size distribution for N4W was similar to that for the 100% clay and no pores larger than around 1.5 μm were present. Thus, nepheline syenite at 4 wt% is shown to have small but measurable effects on the thermal behaviour of brick clays. However, under the conditions used it did not result in significant changes in volumetric shrinkage or water absorption which would have indicated that lower temperatures or times could be used in production.

3. Whether the properties of clays with either the C4W or the N4W compositions are suitable for the production of bricks for construction purposes requires further testing but the observed reductions in the shrinkage onset temperature and the greater extent of the shrinkage around 1000 °C for nepheline syenite may be useful phenomena to exploit further in order to reduce energy use and emissions, especially since both these materials can be sourced as wastes from other processes, leading to a further reduction in the environmental burden.

Acknowledgements The authors would like to thank Wienerberger Ltd. for providing the raw material (clay) and funding this research with Sheffield Hallam University.

References

1. Vogl, Ceramic roadmap 2050—Whitewares' contribution **90**, pp. E34–E38 (2013)
2. D.G. WSP PB, Decarbonisation, Industrial Roadmap, Energy Efficiency Plan, Action, no. October, 2017
3. N.V. Boltakova, G.R. Faseeva, R.R. Kabirov, R.M. Nafikov, Y.A. Zakharov, Utilization of inorganic industrial wastes in producing construction ceramics. Review of Russian experience for the years 2000–2015. Waste Manag. **60**, 230–246 (2017). https://doi.org/10.1016/j.wasman.2016.11.008
4. A.B. Dondi, M. Marsigli, Fabbri, Review recycling of industrial and urban wastes in brick Production. **13**(3), 218–225 (1997)
5. A.K. Abdurakhmanov, A.M. Éminov, G.N. Maslennikova, Stages of ceramic structure formation in the presence of additives. Glass Ceramics (English Translation of Steklo i Keramika) **57**(9–10), 354–356 (2000). https://doi.org/10.1023/A:1007150606044

6. E. Lewicka, Conditions of the feldspathic raw materials supply from domestic and foreign sources in Poland. Gospodarka Surowcami Mineralnymi/Mineral Resources Management **26**(4), 5–19 (2010)

7. J. Pranckevičiene, V. Balkevičius, A.A. Špokauskas, Investigations on properties of sintered ceramics out of low-melting illite clay and additive of fine-dispersed nepheline syenite. Medziagotyra **16**(3), 231–235 (2010)

8. M.U. Rehman, M. Ahmad, K. Rashid, Influence of fluxing oxides from waste on the production and physico-mechanical properties of fired clay brick: a review. J. Build. Eng. **27**, 100965 (2020). https://doi.org/10.1016/j.jobe.2019.100965

9. J. Everhart, J. Lawson Felder, Process for fast-fire ceramic tile using nepheline syenite and clay (1972)

10. R.L. Frost, R. Scholz, X. Ruan, R.M.F. Lima, Thermal analysis and infrared emission spectroscopy of the borate mineral colemanite ($CaB_3O_4(OH)_3 \cdot H_2O$): Implications for thermal stability. J. Therm. Anal. Calorim. **124**(1), 131–135 (2016). https://doi.org/10.1007/s10973-015-5128-5

11. A. Yamık, İ. Bentli, C. Karagüzel, M. Çınar, B. Cengiz, The Application of Colemanite Addition to Floor Tile Glazes, pp. 753–756 (2001)

12. S. Akpinar, A. Evcin, Y. Ozdemir, Effect of calcined colemanite additions on properties of hard porcelain body. Ceram. Int. **43**(11), 8364–8371 (2017). https://doi.org/10.1016/j.ceramint.2017.03.178

13. A.M.T. Bell et al., X-ray Fluorescence Analysis of Feldspars and Silicate Glass Effects of Melting Time on Fused Bead Consistency and Volatilisation, pp. 1–17 (2020)

14. A. Rusen, Investigation of structural behavior of colemanite depending on temperature. Revista Romana de Materiale/Romanian J. Mater. **48**(2), 245–250 (2018)

15. R.G. Frizzo, A. Zaccaron, V. de Souza Nandi, A.M. Bernardin, Pyroplasticity on porcelain tiles of the albite-potassium feldspar-kaolin system: a mixture design analysis. J. Build. Eng. **31**, 101432 (2020). https://doi.org/10.1016/j.jobe.2020.101432

16. I. Waclawska, L. Stoch, J. Paulik, F. Paulik, Thermal decomposition of colemanite. Thermochimica Acta **126**(C), 307–318 (1988). https://doi.org/10.1016/0040-6031(88)87276-9

17. L. Beddiar, F. Sahnoune, M. Heraiz, D. Redaoui, Thermal transformation of fired clay ceramics by dilatometric analysis. Acta Phys. Pol. A **134**(1), 86–89 (2018). https://doi.org/10.12693/APhysPolA.134.86

18. F.C. Kracek, The polymorphism of sodium sulfate (1927). https://doi.org/10.1021/j150303a001

19. N. Ediz, A. Yurdakul, Characterization of porcelain tile bodies with colemanite waste added as a new sintering agent. J. Ceramic Process. Res. **10**(4), 414–422 (2009)

20. H. Giesche, Mercury porosimetry: a general (practical) overview. Part. Part. Syst. Charact. **23**(1), 9–19 (2006). https://doi.org/10.1002/ppsc.200601009

21. K.K. Aligizaki, Pore Structure of Cement-based Materials: Testing, Interpretation and Requirements. Crc Press (2005)

22. P.S. Nicholson, W.A. Ross, Kinetics of oxidation of natural organic material in clays. J. Am. Ceram. Soc. **53**(3), 154–158 (1970). https://doi.org/10.1111/j.1151-2916.1970.tb12058.x

Chapter 33
Mechanical Strength of Poly Nanofiber Patch Under a Biaxial Tensile Loading

Elif Sensoy and Mahmoud Chizari

Abstract Most conventional material testing apparatuses are unable to assess poly-nanofibers sheets in biaxial directions. This study reports the design and prototyping of a biaxial tensile apparatus which can measure the mechanical property of a poly nanofibers patch. Several samples were assessed using the designed biaxial tensile testing machine and results recorded. Function of the apparatus was validated versus convention methods and outcome confirmed that it is accurate and reliable for testing poly nanofibers patch.

Keywords Biaxial tensile testing · Nanofibers · Electrospinning · Poly lactic acid · Mechanical properties

33.1 Introduction

Nanofibers are defined as a material fabricated with nanometres fibres. Development and research on the nanofibers have become the interest of many industries [1]. The aim of this project is to design and develop a machine to measure the mechanical properties of poly-nanofiber sheets. This has been of interest to many different sectors including; bio medical engineering, textiles, automotive and many more. The interest in the topic has risen due to a rise in the different application of nanofibers [2]. This report will be focusing on testing the mechanical properties of poly-nanofiber patches for use of biomedical engineering and tissue engineering. Currently there isn't a very robust or reliable method of measuring the mechanical strength of poly-nanofibers which causes issues when developing nanofiber patches for uses where they will be required to have high levels of mechanical strength. The current Biaxial tensile machines available for use are not sufficient for testing poly nanofibers, as these machines have tolerances only suitable for testing conventional materials such as metals and some composites. As the demand for poly-nanofibers increase, suitable methods of testing are required, this report will show the design and development

E. Sensoy (✉) · M. Chizari
University of Hertfordshire, College Lane, Hatfield AL10 9AB, UK
e-mail: Elifsensoy1997@gmail.com

© The Author(s) 2021
I. Mporas et al. (eds.), *Energy and Sustainable Futures*, Springer Proceedings in Energy,
https://doi.org/10.1007/978-3-030-63916-7_33

process and the testing process of an apparatus which is suitable for testing the mechanical properties of poly-nanofibers. The design considered must be able to test specimen with a size of 30×30 mm. The measurement should be in biaxial directions and the system should be able to distribute enforced load evenly across the specimen in cross directions. The grippers holding the specimen in place, must not damage the nanofibers specimen, whilst also being able to provide enough friction to be able to hold the specimen in place and distribute the enforced load evenly. The system should be able to measure the displacement and stretch displacement of the material. It should also let the operator observe the mode of failure on the specimen. The system must be safe to use and handle. The measured data in form of force and displacement must be transferable to a computer. The results must be available for transferring into other mathematical analysis software, i.e. Microsoft excel. Furthermore, the grippers must come with sensors to consider necessary friction on the specimens.

33.2 Design Methodology

Once the technical design specifications were thoroughly analysed, complying design concepts were developed. Each concept was considered with conjunction to the time frame available, cost, and environmental impact. Different mechanism design options were developed and thought about, the design concepts were compared, and some elements were used in conjunction with another. The most feasible and well performing designs were taken forward. Nearly all the design concepts featured both electronic and mechanical elements. This unity provides accuracy, precision and practicality during both production and the use of the design. During design development, each design was thought about carefully, they were constantly changed to be able to improve the design even more. There were several different iterations of the same design with some elements changed, to be able to improve the overall outcome. The goal was to create a flexible design with the ability to change the parts practically during and after the design process [3]. It was important to also find a reliable method of transferring data from the Arduino software to Microsoft Excel, after thorough research it was concluded that Tera-Term would be used and integrated into the Arduino programme, so that the data yielded from the tests could be viewed in an Microsoft Excel document. The tests are to be completed 10 times to ensure that the experiment allows for accurate results to be yielded and to ensure that the test is repeatable.

The device was calibrated through Arduino programming and to ensure that the calibration done through programming software was accurate and successful, a ruler was used and attached next to the arm, to check if the measurement from the displacement mechanism made up of a small pulley mechanism. The load cells were calibrated using a 100 grams weight which was previously measured several times, then the weights were put on the load cell, to ensure that the load cells were recognising the correct values of load and force applied. This was done to ensure the repeatability of

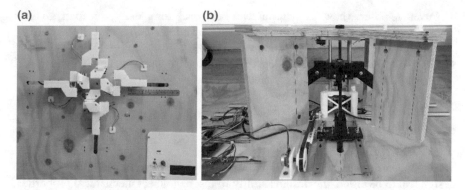

Fig. 33.1 Front view of apparatus (a), back of apparatus showing the operating mechanism

the experiment. The values measured from both the load cells and the displacement mechanism are all displayed on the LCD display on the control panel (Fig. 33.1). Following the prototyping of the machine, it was ready for a practical evaluation. To test a sample following stages were performed:

- Connect the apparatus to mains power socket and to the PC via the USB port on the control panel of the device.
- Turn on the power of the motor using the motor power switch located on the control panel, then reset the arm position and sensors by pressing load cell reset and displacement reset buttons also located on control panel.
- Ensure the specimen is ready to place and secure into the apparatus grippers. Lift the latches of each gripper and carefully place the specimen into the grippers. Close all lids and screw latches into place with the help of a screwdriver, for extra support hold the arm from the bottom to make sure the screws are in place properly.
- On the PC launch TeraTerm, once it has launched, a page called TeraTerm: New connection will come up, now select the serial option, and select the USB port used to connect the apparatus to the PC. Click OK.
- Once step 4 is complete, a window named COMX-Tera Term VT will show up on your screen. In the top, left hand corner click file, then click on log. Give the data sheet a name and add .csv to the end of the document. This will enable the data sheet to be opened as a Microsoft Excel document, once the document is named using the correct file extension click save.
- Now select file again, and under file click show dialog box, a window called TeraTerm Log will show up on the screen. Click pause so the data logging stops at that moment in time.
- Ensure that everything is secured in place and ready for testing, then click start and rotate the potentiometer located on the control panel. This will start the test and exert the load on the specimen by pulling it in all 4 directions.

- Ensure to observe how the material tears, and once completely torn apart, go back to the PC and click pause on the TeraTerm Log window. Close the Tera Term log and COMX- Tera Term VT Windows.
- Go to the file where the data sheet saved from the previous steps, the file will automatically launch as an excel sheet.

33.3   Test Results

When the design and prototyping process was complete, the performance of the design had to be tested, the designs ability to test nanofibers mechanical properties, the accuracy and reliability of the results provided through the testing achieved using the device. The data yielded from testing are all within range of the average. Thus, proving the accuracy and reliability of the device. The results of the experiments complete can be seen in Figs. 33.2 and 33.3.

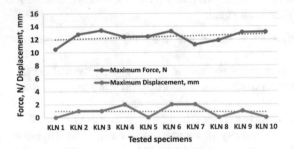

Testes samples	Max. Force, N	Max. Displacement, mm
KLN 1	10.515	0
KLN 2	12.816	1
KLN 3	13.408	1
KLN 4	12.421	2
KLN 5	12.474	0
KLN 6	13.269	2
KLN 7	11.212	2
KLN 8	11.887	0
KLN 9	13.099	1
KLN 10	13.159	0
Avg.	12.426	0.9

Fig. 33.2 Graph of maximum load and displacement and table of measured load and displacement

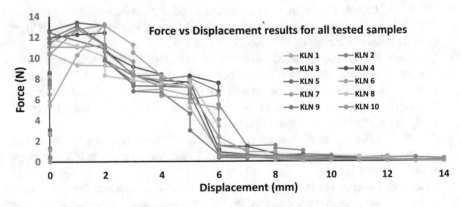

Fig. 33.3 Test results of 10 samples; Force (N) vs Displacement (mm)

33.4   Discussion

The data yielded from tests completed, have illustrated the reliability and the accuracy of the device prototyped. The results produced from all 10 experiments conducted are all within the range of the average of the results yielded from the testing calculated in Fig. 33.2 as 0.9 mm for displacement and 12.426 N for maximum force. This shows that the experiment is repeatable, this also shows that the results obtained through testing and experimentation are reliable and accurate.

As can be seen in Figs. 33.2 and 33.3 some results can be classified as outliers, looking at the maximum forces in the Fig. 33.2 it can be seen that the sample, KLN 7, has a lower value than the other maximum forces yielded from the other test iterations. This could be due to different parameters such as, microtears in ply of tissue used, micro cracks due to production procedures and handling techniques e.g. fold lines. There are, of course, many parameters which could have caused inaccuracies in the results obtained, for example the homogeneity of the fibres in the Kleenex facial tissues (Kimberly-Clark Worldwide, US), and the creases caused due to the packaging of the tissues. This can also be seen from the graphs obtained through the results obtained. There are some results which can be classified as outliers, these outliers which may have been caused by the parameters discussed previously. The reason for using Kleenex tissues as test specimen is due to the ongoing adverse circumstance and unavailability of resources.

There is positive correlation between the force applied and the amount of displacement that takes place before the material deforms this can be seen in Fig. 33.3, where the materials displacement increases as the force exerted increases until the material fractures. The results which can be seen in Figs. 33.2 and 33.3 were obtained by keeping all parameters in control and steady where possible, to ensure the results yielded illustrated clearly how accurate the device is. An interesting observation made from the results obtained through testing is that the specimen undergoing testing, behaves similarly to a brittle material, meaning that the load undergone increases rapidly but no displacement is recorded, then the maximum force is reached which causes the material to fracture, then the specimen rapidly starts tearing and deforming (Fig. 33.4).

33.5   Conclusion

This study documents the design, development and prototyping of a biaxial tensile testing machine produced for the purposes of testing poly nanofiber patches. The design and prototype were validated through a series of comprehensive sample testing and data processing. The data yielded from testing were analysed and it was concluded that the result generated from the experiment were accurate and reliable, also that the design developed, and prototype was sufficient and performed well for the purposes of testing poly nanofibers.

Fig. 33.4 Circuitry located
behind control panel

References

1. D. Alejandro Arellano Escarpita. Biaxial Tensile Strength Characterization of Textile Composite
   Materials. Tlalpan, México D.F., Mexico: Mechatronics Engineering Department, Instituto
   Tecnológico y de Estudios Superiores de Monterrey, Campus Ciudad de México, Col. Ejidos de
   Huipulco (2012)
2. Injection rate may effect morphology of nanofibres made by electrospinning. S. Rafiei, A.
   Nourani, M.H. Abedini, F. Manshae, M. Mohseni, M. Chizari. Istanbul, Turkey: 5th interna-
   tional conference on advances in mechanical engineering, Yıldız Technical University, 17–19
   December 2019
3. R. Casasola, N.L. Thomas, A. Trybala, S. Georgiadou, *Electrospun Poly Lactic Acid Fibres:
   Effect of Different Solvent on Fibre Morphology and Diameter* (Loughborough University,
   Loughborough, Leicestershire, 2015)

Chapter 34
Damage Characterisation in Composite Laminates Using Vibro-Acoustic Technique

Kristian Gjerrestad Andersen, Gbanaibolou Jombo,
Sikiru Oluwarotimi Ismail, Yong Kang Chen, Hom Nath Dhakal,
and Yu Zhang

Abstract The need to characterise in-service damage in composite structures is increasingly becoming important as composites find higher utilisation in wind turbines, aerospace, automotive, marine, among others. This paper investigates the feasibility of simplifying the conventional acousto-ultrasonic technique set-up for quick and economic one-sided in-service inspection of composite structures. Acousto-ultrasonic technique refers to the approach of using ultrasonic transducer for local excitation while sensing the material response with an acoustic emission sensor. However, this involves transducers with several auxiliaries. The approach proposed herewith, referred to as vibro-acoustic testing, involves a low level of vibration impact excitation and acoustic emission sensing for damage characterisation. To test the robustness of this approach, first, a quasi-static test was carried out to impute low-velocity impact damage on three groups of test samples with different ply stacking sequences. Next, the vibro-acoustic testing was performed on all test samples with the acoustic emission response for the samples acquired. Using the acoustic emission test sample response for all groups, the stress wave factor was determined using the peak voltage stress wave factor method. The stress wave factor results showed an inverse correlation between the level of impact damage and stress wave factor across all the test sample groups. This corresponds with what has been reported in literature for acousto-ultrasonic technique; thus demonstrating the robustness of the proposed vibro-acoustic set-up. Structural health monitoring, impact damage, acousto-ultrasonic testing, non-destructive testing.

K. G. Andersen · G. Jombo (✉) · S. O. Ismail · Y. K. Chen
Centre for Engineering Research, University of Hertfordshire, Hatfield, UK
e-mail: g.jombo@herts.ac.uk

H. N. Dhakal
School of Mechanical and Design Engineering, University of Portsmouth, Portsmouth, UK

Y. Zhang
Department of Aeronautical and Automotive Engineering, Loughborough University,
Loughborough, UK

© The Author(s) 2021
I. Mporas et al. (eds.), *Energy and Sustainable Futures*, Springer Proceedings in Energy,
https://doi.org/10.1007/978-3-030-63916-7_34

34.1 Introduction

Composite materials are increasingly been used in various industries such as energy, automotive, aerospace, marine, to mention but a few, as critical structural elements. This increasing use of composite materials necessitates the need for characterising in-service damage. Composite is formed by the combination of two materials with different physical and chemical properties. They are used for structural applications due to their light weight, high specific strength/stiffness and good resistance to a corrosive environment. Damage to composite structures can be invisible to the naked eye and non-destructive testing (NDT) techniques such as ultrasound, pulsed thermography and acoustic emission are often used to detect faults, such as delamination, matrix cracking, fibre-matrix de-bonding, fibre breakage and matrix porosity [1–3].

More also, for the past few decades, accurate characterisation of damage in fibre-based composites has been an area of active research [3–5]. Advancing this goal, this paper investigates the characterisation of impact damage in fibre reinforced polymer (FRP) composites by using a variant of acousto-ultrasonic technique (AUT).

Creating a viable procedure for damage characterisation that generally works for various fibre-based composites has proven to be challenging, because of the anisotropic nature of fibre-based composites. Moreover, impact response to composite structures can cause damage such as delamination, matrix cracking and de-bonding in the laminate.

AUT was first proposed by Alex Vary as an NDT technique for evaluating the inter-laminar shear strength of fibre composites [6]. AUT combines acoustic emission (AE) methodology and ultrasonic simulation of stress waves. In order to quantify variations in the mechanical properties, Vary proposed a measurement parameter called the stress wave factor (SWF). SWF is a descriptive parameter that correlates with the material properties of composite materials. Contrary to traditional acoustic emission technique that requires the material to be under stress, AUT does not.

In recent times, AUT has found application in the following areas for NDT of composite structures: damage detection and severity quantification [7], material property correlation of fibre-based composites [8] and naturally occurring composites [9].

Importantly, this paper addresses a different yet important question, "can the set-up for AUT be simplified for quick and economic one-sided in-service composite structure inspection?" To answer this question, a variant set-up of AUT is proposed and explored. This approach involves a low level vibration impact excitation and acoustic emission sensing for damage characterisation in composite structures.

34.2   Experimental

34.2.1   Sample Preparation

The test samples were made from an epoxy impregnated carbon fibre laminates (prepreg) with stacking sequence, as shown in Table 34.1. Hand lay-up method was used to prepare the test samples, in addition to autoclave curing to improve their mechanical properties. The curing cycle in the autoclave involved temperature ramp-up stage from ambient of 20 to 121 °C at a rate of 1 °C/min, followed by a dwell stage to maintain the temperature constant at 121 °C for two hours; and then ramp-down stage, with temperature being reduced to ambient temperature of 20 °C. The internal pressure of the autoclave was maintained at 106 kPa for the entire temperature cycling operation.

34.2.2   Testing

34.2.2.1   Low Velocity Impact Damage (Quasi-static Testing)

Composite laminate structures are very prone to low-velocity impact damage. Although, the effects of low-velocity impact damage are barely visible on the laminate surfaces, there can be extensive sub-surface damage in the form of matrix cracking, delamination and fibre failure, leading to a reduction in residual strength after an impact [10].

To experimentally simulate the effects of low-velocity impact damage in the test samples, a low-velocity impact experiment was performed on a Tinius Olsen Model 25ST Universal Benchtop Tester, as shown in Fig. 34.1, with the test samples in Table 34.1. The set-up consists of a hemispherical indenter with diameter of 25.4 mm and four toggle clamps attached to a steel plate with an open middle slot to enable

Table 34.1   Test Sample stacking sequence

Test Sample	Material	Stacking sequence	Dimensions (mm)	Laminate type	Number of samples
A	FibreDUX 6268C-HTA 12 K	[90/± 45/0]$_s$	150 × 135 × 2	Quasi-Isotropic	4
B	FibreDUX 6268C-HTA 12 K	[90/0/± 45]$_s$	150 × 135 × 2	Quasi-Isotropic	4
C	FibreDUX 6268C-HTA 12 K	[90/0]$_{2s}$	150 × 135 × 2	Quasi-Isotropic	4

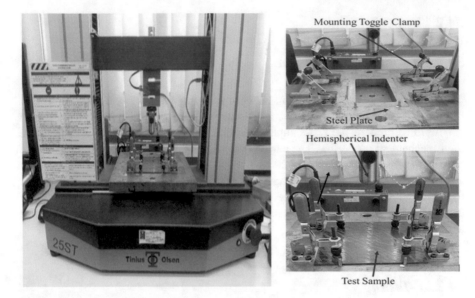

Fig. 34.1 Low velocity impact test set-up

the impacted zone to deform or fracture during the impact testing. The test samples were designated with numbers 0–3, where 0 represented non-impacted samples and other 3 samples were impacted with a velocity of 2 mm/s with increasing maximum impact forces of 2.00, 2.25 and 2.50 kN, respectively.

34.2.2.2 Vibro-Acoustic Testing

The conventional AUT set-up consists of two piezoelectric transducers, as shown in Fig. 34.2. The transmitting transducer is an ultrasonic emitter, while the receiving transducer is an AE sensor [11].

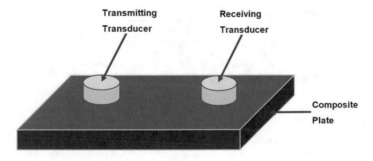

Fig. 34.2 Conventional AUT set-up

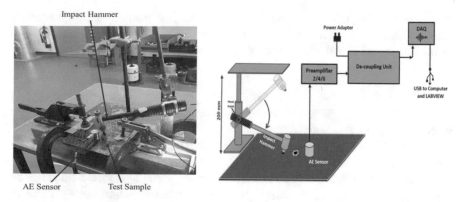

Fig. 34.3 Vibro-Acoustic test set-up

In order to explore the feasibility of simplifying the conventional AUT set-up for quick and cost effective one-sided in-service inspection of composite structures, a vibro-acoustic set-up shown in Fig. 34.3 is proposed. It consists of an impact hammer clamped to a rig with a pivot mechanism and a stopper of maximum height of 200 mm to ensure a repeatable low energy level excitation; nearly 59 N. R15α AE sensor from Physical Acoustics has a resonant frequency of 150 kHz, was coupled to a 2/4/6 preamplifier with a gain of 60 dB selected. Impact hammer of model 086-B02 from PCB Piezotronics was used. The NI USB-6361 data acquisition system from National Instruments with maximum sampling frequency of 2 MHz was also used. Total sample acquisition time of 30 s with excitation occurring after 10 s. To ensure the test samples did not move while being excited, they are held down with one toggle clamp on either side with cardboard-patches to prevent damage. For repeatability and also to minimise signal attenuation by the composite test samples, the AE sensor was placed approximately 20 mm from the impact zone, while the hammer hit was done approximately 15 mm from the impact zone. This sensor spacing (30–50 mm) corresponds with what has been reported to minimise the effect of signal attenuation [7, 8].

34.3   Results and Discussion

In an AUT domain, a commonly used approach for characterising a damage in fibre reinforced composite laminate is by calculating the stress wave factor (SWF). SWF quantifies the attenuation of the material to the induced stress wave. A low SWF corresponds to a region with higher attenuation, due to sub-surface damage and a relatively high SWF indicates a region of lower attenuation.

SWF can be determined with any of the following methods: peak voltage SWF method, ringdown SWF method, weighted ringdown SWF method and energy integral SWF method [12]. Each method for SWF resulted to different values for the parameter. For this investigation, the peak voltage method was applied. The peak voltage SWF method which assumed an inverse relationship between the peak-voltage and signal attenuation, due to damage in material is represented as Eq. (34.1).

$$SWF = V_{max} \tag{34.1}$$

where V_{max} is the peak voltage.

Figures 34.4 and 34.5 show the time domain AE response and SWF analysis from the vibro-acoustic tests.

Figure 34.4 depicts the time domain response of the AE signal and a trend can be observed: the peak induced test sample response from the excitation was significantly reduced, showing difference between the unimpacted samples (A0, B0 and C0) and the impacted samples (A1–A3, B1–B3 and C1-C3). Figure 34.5 shows a significant reduction in SWF between the unimpacted samples (A0, B0 and C0) and (A1, B1 and C1), which was subjected to 2 kN quasi-static load. This response suggests damage progression in the composite laminates. However, in Fig. 34.5, there was only a minor difference in SWF between (A2, B2, C2) and (A3, B3 and C3), at these points there were already significant fibre breakage and matrix cracking present in the impacted area. The result obtained agree with similar published ones [7].

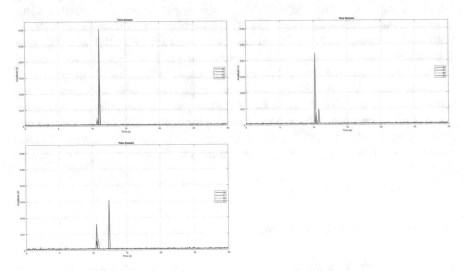

Fig. 34.4 Time domain AE response for test samples A(0–3), B(0–3) and C(0–3)

Fig. 34.5 Peak voltage SWF analysis for test samples A(0–3), B(0–3) and C(0–3)

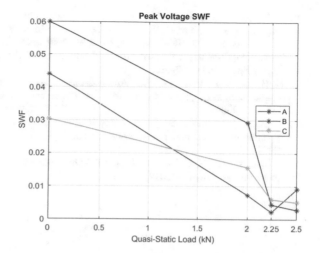

34.4   Conclusions

This investigation explored the feasibility of simplifying the conventional set-up of AUT testing for quick and economic one-sided composite laminate inspection. Hence, the following inferences have been deduced from the study.

- The vibro-acoustic set-up proposed is viable and robust for damage detection and characterisation in fibre reinforced polymer composite laminate.
- Furthermore, the SWF analysis plot showed an inverse correlation relationship between the level of impact damage and SWF across all the test sample groups. This agrees with what has been reported in the literature for AUT [7].

References

1. Q. Shen, M. Omar, S. Dongri, Ultrasonic NDE Techniques for impact damage inspection on CFRP laminates. J. Mater. Sci. Res. **1**(1), 1–16 (2012)
2. Z. Wang, J. Zhu, G. Tian, F. Ciampa, Comparative analysis of eddy current pulsed thermography and long pulse thermography for damage detection in metals and composites. NDT E Int. **107**, 1–10 (2019)
3. M. Saeedifar, D. Zarouchas, Damage Characterization of Laminated Composites Using Acoustic Emission: A Review. Compos. Part B Eng. **195**, 1–21 (2020)
4. Z. Su, L. Ye, Y. Lu, Guided lamb waves for identification of damage in composite structures: a review. J. Sound Vib. **295**(3), 753–780 (2006)
5. M. Mitra, S. Gopalakrishnan, Guided wave based structural health monitoring: a review. Smart Mater. Struct. **25**(5), 1–27 (2016)
6. A. Vary, Acousto-ultrasonic characterization of fiber reinforced composites. Mater. Eval. **40**, 650–654 (1982)
7. L.S.S. Pillarisetti, R. Talreja, On quantifying damage severity in composite materials by an ultrasonic method. Compos. Struct. **216**, 213–221 (2019)

8. C. Barile, C. Casavola, G. Pappalettera, P.K. Vimalathithan, Acousto-ultrasonic evaluation of interlaminar strength on CFRP laminates. Compos. Struct. **208**, 796–805 (2019)
9. H.L.M. dos Reis, F.C. Beall, M.J. Chica, D.W. Caster, Nondestructive evaluation of adhesive bond strength of finger joints in structural lumber using the acousto-ultrasonic approach. J. Acous. Emiss. **9**(3), 197–202 (1992)
10. H.-P. Cui, W.-D. Wen, H.-T. Cui, An integrated method for predicting damage and residual tensile strength of composite laminates under low velocity impact. Comput. Struct. **87**(7), 456–466 (2009)
11. T. Liu, S. Kitipornchai, M. Veidt, Analysis of acousto-ultrasonic characteristics for contact-type transducers coupled to composite laminated plates. Int. J. Mech. Sci. **43**(6), 1441–1456 (2001)
12. S. Mareeswaran, T. Sasikumar, The acousto-ultrasonic technique: a review. Int. J. Mech. Eng. Technol. **8**(6), 418–434 (2017)

Printed in the United States
by Baker & Taylor Publisher Services